Common Misconceptions in Mathematics

Strategies to Correct Them

Bobby Ojose

UNIVERSITY PRESS OF AMERICA,® INC.
Lanham • Boulder • New York • Toronto • Plymouth, UK

Copyright © 2015 by
University Press of America,® Inc.
4501 Forbes Boulevard
Suite 200
Lanham, Maryland 20706
UPA Acquisitions Department (301) 459-3366

Unit A. Whitacre Mews, 26-34 Stannary Street,
London SE11 4AB, United Kingdom

All rights reserved
Printed in the United States of America
British Library Cataloging in Publication Information Available

Library of Congress Control Number: 2012936170
ISBN: 978-0-7618-5885-0 (paperback : alk. paper)
eISBN: 978-0-7618-5886-7

∞™ The paper used in this publication meets the minimum
requirements of American National Standard for Information
Sciences—Permanence of Paper for Printed Library Materials,
ANSI Z39.48-1992

This book is dedicated to my children from whom some of the misconception ideas emanated: Teriji, Ejiro, Ese, Kessiena, and Elohor. Also, to students, teachers, and math coaches that I have worked with in my career as a mathematics educator.

Contents

Acknowledgments — ix

Introduction — xi
   The Purpose of the Book
   Issues with Misconceptions
      What Are Misconceptions in Mathematics?
      How Do Misconceptions Come About?
      Why Is It Important to Correct Misconceptions?

PART ONE: ARITHMETIC — 1

Misconception 1: Addition Sentence — 3

Misconception 2: Subtraction of Whole Numbers — 6

Misconception 3: Addition of Fractions — 8

Misconception 4: Subtraction of Fractions — 10

Misconception 5: Rounding Decimals — 13

Misconception 6: Comparing Decimals — 15

Misconception 7: Multiplying Decimals — 17

Misconception 8: More on Multiplying Decimals — 19

Misconception 9: Division of Decimals — 22

Misconception 10: Percent Problems — 25

Misconception 11: Division by a Fraction — 28

Misconception 12: Ordering Fractions 30

Misconception 13: Least Common Multiple (LCM) 32

Misconception 14: Addition of Decimal Numbers 34

Misconception 15: Subtraction of Integers 37

Misconception 16: Converting Linear Units 39

Misconception 17: Power to a Base 42

Misconception 18: Order of Operations I 45

Misconception 19: Order of Operations II 47

Misconception 20: Simplifying Square Roots 49

Misconception 21: Comparing Negative Numbers 51

Misconception 22: Addition of Negative Integers 53

Misconception 23: Scientific Notation 56

Misconception 24: Proportional Reasoning 58

Misconception 25: Time Problem 61

PART TWO: ALGEBRA 63

Misconception 26: Dividing Rational Expressions 65

Misconception 27: Adding Rational Expressions 68

Misconception 28: Adding Unlike Terms 71

Misconception 29: Adding Like Terms 74

Misconception 30: Distributive Property 77

Misconception 31: Writing a Variable Expression 80

Misconception 32: Simplifying a Variable Expression 83

Misconception 33: Factoring 86

Misconception 34: Exponents Addition 89

Misconception 35: Zero Exponents 93

Misconception 36: Solving Equation by Addition and Subtraction ... 97

Misconception 37: Solving Equation by Division and Multiplication 100

Misconception 38: Fractional Equations	103
Misconception 39: One-Step Inequality	106
Misconception 40: Absolute Value	109
Misconception 41: Operations with Radical Expressions	112
Misconception 42: Simplifying Polynomials	115
Misconception 43: Systems of Equations	118
Conclusion	122
Appendix A: List of Manipulatives and Their Uses	124
Appendix B: Teaching Standards	125
References	126
About the Author	135

Acknowledgments

My vote of thanks goes to the reviewers that helped a great deal in shaping the content and structure of the book. They include Dr. Janet Beery, a Mathematics Professor at the University of Redlands, California; Dr. Ramakrishnan Menon, a Mathematics Education Professor at George Gwinnett College, Georgia; Dr. Rong-Ji Chen, an Associate Professor of Mathematics Education at California State University, San Marcos; and Dr. Fred Uy, an Associate Professor of Mathematics Education at California State University, Los Angeles. Thanks to Catherine Walker and Kimberly Perna of the Information Technology Service (ITS) department at the University of Redlands. They were patient in formatting some of the figures and I am quite grateful. I wish to also recognize those that have influenced me as a professor at the University of Redlands where I started my college teaching career: Dr. Ron Morgan for support and kindness; Dr. Jose Lalas for the informal mentorship and the constant push for me to publish; Dr. Chris Hunt for being such a friendly person; Dr. Phil Mirci for opening me up to the doctoral program; and Dr. Jim Pick for the scholarship networking. The staff of the School of Education at the University of Redlands also deserve many thanks, especially Colleen Queseda, Office Manager. At the Beeghly College of Education in Youngstown State University, I would like to thank the Dean, Dr. Charles Howell for granting my first research course release request. Last but not least, I would like to thank Dr. Marcia Matanin for opening up the avenue for me to be a program coordinator of the Middle Childhood Education program.

Introduction

THE PURPOSE OF THIS BOOK

This book should be used by pre-service and in-service teachers whose present or future students would come to their class with misconceptions. Regardless of the source of a misconception, it is important for the teacher to detect them and plan instruction that addresses them. For each identified misconception in this book, there is a discussion of what the student is doing wrong or fails to do right. Also, in discussing the misconception, the sources of the problem are pointed out because for a correction technique to be effective, teachers need to appreciate the source of the problem. Next, the methods that teachers could use to help students overcome the misconception are highlighted. These suggested strategies are in most cases nontraditional. They include the use of models, discussions, group activities, etc.

The specific tasks or teaching strategies that a teacher could use in dealing with each identified misconception are discussed. In some instances, the discussion on methodology revolves around the things that a teacher might want to also avoid. The research notes are for teachers interested in knowing more about studies that have been carried out with the particular mathematics concepts or issues related to mathematics education in general. The brief summary of research findings and comments are provided with the intent that teachers who may not have the time to read an entire cited work can have a quick glimpse of what researchers are saying about the topic. It is my hope that the knowledge of the existence of misconceptions in mathematics and conscious efforts on the part of teachers to correct them would positively impact students of mathematics. Therefore, the book serves as a starting point for teachers willing to make a difference with issues relating to mathematics misconceptions.

ISSUES WITH MISCONCEPTIONS

What Are Misconceptions in Mathematics?

Misconceptions are misunderstandings and misinterpretations based on incorrect meanings. They are due to 'naïve theories' that impede rational reasoning of students. Misconceptions take various forms. For example, a correct understanding of money embodies the value of coin currency as non-related to its size. But at the pre-kindergarten level, children hold a core misconception about money and the value of coins. Students think nickels are more valuable than dimes because nickels are bigger. Some elementary and even middle school students thinks that 1/4 is larger than 1/2 because 4 is greater than 2. Another example is in multiplication of numbers. A correct understanding of multiplication includes the knowledge that multiplication does not always increase numbers. But students have a misconception that multiplication always increases a number. This impedes students' learning of the multiplication of a positive number by a fraction less than one, such as $1/2 \times 14 = 7$.

Mathematics textbooks do not treat the issue of misconceptions directly. Therefore, teachers' dependency on these texts for instruction further minimizes the misconception issue. Because some of these misconceptions occur over and again in each grade, it is important that concerted efforts are made by mathematics educators at correcting them. Teachers should be aware of the existence of misconception issues in mathematics. Awareness coupled with development of effective corrective strategies should help in exposing students to the correct way of thinking about mathematical concepts.

How Do Misconceptions Come About?

Students generally make errors that take two forms: conceptual errors and execution errors. Conceptual errors are related to lack of understanding. Examples are *structural* which involves failure to appreciate relationships in the problem or to grasp some principles essential for its solution or *arbitrary* which involves lack of loyalty to the givens of the problem under the influence of previous experiences. On the other hand, execution errors happens when an attempt to carry out some procedure breaks down or is only partially executed. Executive errors arise not from failure to understand how the problem should be tackled, but in some failure to actually carrying out the manipulations required.

It is also important to mention that misconceptions exist in part because of students' overriding need to make sense of the instruction that they receive. For example, the rules of adding fractions with like and unlike denominators are quite different. Moving from adding fractions with like denominators to adding fractions with unlike denominators requires students to make sense of the different scenarios and make adjustments. The transition often creates cognitive conflicts and dissonance for students because the process requires unlearning what has been previously learned.

In some other instances, misconceptions can be more enduring – students predictably err across a range of problems for considerable length of time. This usually happens because students probably had memorized rules and relied on rote learning for some time. When the time comes to solve a problem, the memorized rules may not be appropriate for the task at hand as it requires conceptual knowledge. Without guidance, a student might apply the inappropriate rule to the current problem and thereby resulting in more mathematical misunderstanding.

One cannot help but understand how these misconceptions manifest based on the nature of mathematics. The nature of mathematics is such that the rules keep changing from one concept to another and from one mathematical operation to another. For example, when decimals are introduced with addition, $0.4 + 0.7$ is 1.1 (one decimal place), but with multiplication of decimals, 0.4×0.7 is 0.28 (two decimal places). The discrepancy from addition operation to multiplication operation in decimals could be a reason for students to have misconceptions. Another dimension related to the nature of mathematics is that certain misconceived methods and errors in calculation could actually lead to right answers, probably the more reason why students seem to hang on to them. For example, if $1/9$ is divided by $1/3$, the answer is $1/3$. Students could erroneously divide out the numerators to get 1 and also divide out the denominators to get 3 and thereby arriving at the right answer of $1/3$ (through the wrong method). When this kind of situation happens, the onus is on the classroom teacher to correct the error. In general, knowing the nature of an error and the source of an error could help teachers fathom ways of planning appropriate instruction that is beneficial to students.

Why Is It Important to Correct Misconceptions?

Understanding issues related to misconceptions is one important step to improving instruction in mathematics. After a teacher detects a misconception with students (be it conceptual or execution), the teacher should device strategies to help students overcome the misconception. As Graeber and Johnson (1991) commented,

It is helpful for teachers to know that misconceptions and "buggy" errors do exist, that errors resulting from misconception or systematic errors do not signify recalcitrance, ignorance, or the inability to learn; that such errors and misconceptions and the faulty reasoning they frequently signal can be exposed; that simple telling does not eradicate students' misconceptions or "bugs" and that there are instructional techniques that seem promising in helping students overcome or control the influence of misconceptions and systematic errors. (pp. 1-2).

As pointed out earlier on, misconceptions would always be experienced by students because of the nature of mathematics. Be as it is, teachers need to be aware of their existence and ensure that misconceptions do not persist with students for a longer period of time. For example, it will be detrimental to a grade four student who is still misplacing decimal points when multiplying decimals to move up to the fifth grade without adequate remedy. Research suggests that misconceptions that persist for years if undetected would negatively affect the future learning of mathematics. For example, Woodward, Baxter, & Howard (1994) pointed out that a continued, superficial understanding of mathematics allows students to apply improper algorithms or repair strategies, eventually resulting in ingrained and deep-seated misconceptions.

As pointed out earlier on, teachers should be sensitive and recognize that students will come to their classroom with misunderstandings and misconceptions. It is imperative for teachers to work towards detecting the existing misconceptions that their students may have and also work hard to correct them. This is to avoid a situation whereby students move from one grade to another with misconceptions. Teachers should therefore acknowledge that students can overcome the misconceptions by planning and consciously providing opportunities for students through effective teaching strategies.

Part One

ARITHMETIC

Part One of the book highlights the misconceptions commonly held by students in general mathematics, especially arithmetic. Teachers and students in grades one to six should find this part of the book useful. It explores misconceptions in number sense topics including concepts like mathematical operations, whole numbers, percents, decimals, and fractions.

Misconception 1

Addition Sentence

Grades: K to 6

Question: Complete the addition sentence by inserting the missing number.
$2 + 8 = \underline{} + 6$

Likely Student Answer: 10

Explanation of Misconception: The student has overgeneralized from experiences with problems in which the equal sign acts as signal to compute. This misconception which is associated with the relational meaning of the equal sign shows that students misconstrue the equal sign as a command to, for example, add the 2 and the 8 to obtain the result of 10. The relational meaning of the equal sign requires students to conceive it as balancing an equation (quantity on one side should be equal to the quantity on the other side).

What Teachers Can Do: The teacher should emphasize the relational meaning of the equal sign. Children must understand that equality is a relationship that expresses the idea that two mathematical expressions hold the same value. Teachers should begin by showing students how 10 is not the answer to the above problem. One way is to have them arrange 2 blocks and 8 blocks as grouping on one side of a mat; and then 10 blocks and 6 blocks as grouping on the other side of the mat. Teacher should ask students if equality holds. With this illustration, students will see practically that the number 10 does not fit in the blank. Next, teacher should ask students to balance the "beam" by making sure that the total number of blocks on the left hand is same as number of blocks on the right side. They should manipulate the right side to get 10 blocks. In other words, *"What should be added to 6 to get 10?"*

It is helpful to engage students in a discussion of the equal sign in the context of specific tasks. Appropriately chosen tasks can (a) provide a focus for student to articulate the ideas, (b) challenge students' conceptions by providing different contexts in which they need to examine the positions they have staked out, and (c) provide a window on children's thinking. Providing answers to true/false number sentences could provide contexts for students to engage with relations involving important mathematical ideas. In other words, just having students respond to true or false simple number sentences can aid their conceptual understanding. For example, students can be made to answer "*True or False*" for the following number sentences: *4 + 6 = 10, 2 + 8 = 10, 10 – 6 = 10, 10 + 6 = 10.*" Answers provided to these questions can act as a spring- bud for highlighting the misconception by students. It is always a good idea to start with number sentences that involve relatively simple calculations with a single number to the right of the equal sign. Once students begin to understand that the equal sign signifies a relation between numbers, it is important to continue to provide number sentences in a variety of forms and not fall back to using only number sentences with the answer coming after the equal sign.

Research Note: Carpenter et al (2003) wrote extensively about misconceptions relating to addition sentence and relational understanding of equality. They showed that fewer than 10 percent of students in any grade gave the correct response to questions of the type $8 + 4 = \_\_\_ + 5$ and that performance did not improve with age. In fact results for the grade six students were actually slightly worse than the results of students in the earlier grades. They hinted that the inappropriate generalizations about the equal sign that students make and often persist in defending are symptomatic of some fundamental limits in their understanding of how mathematical ideas are generated and justified. Students have seen many examples of number sentences of a particular form, and they have overgeneralized from those examples. As they begin to justify conjectures about numbers and number relations, they tend to rely on examples and think that examples alone can prove their case.

One reason why understanding equality as a relationship is important is that a lack of such understanding may be a stumbling block for students transitioning from arithmetic to algebra (Matz, 1982; Falkner, Levi, & Carpenter, 1999; Carpenter, Franke, & Levi, 2003). These researchers have also noted that the understanding of the concept can make student's learning of arithmetic easier and richer. Knuth et al (2006) found a strong positive correlation between middle school students' equal sign understanding and their equation-solving performance. The result showed that even students having no experience with formal algebra (sixth- and seventh- grade students in particular) have a better understanding of how to solve equations when they have a relational understanding of the equal sign. Also, there is a strong positive correlation between equal sign understanding and use of an algebraic strategy among eighth- grade students (who had more experiences with algebraic ideas and symbols) compared to lower grades. Taken together, these findings suggest that understanding the equal sign is important as it is a step toward successful solving of equations. These findings support the importance of explicitly developing students' understanding of the equal sign and equality concepts during their middle school mathematics education.

Misconception 2

Subtraction of Whole Numbers

Grades: 1 to 3

Question: Subtract:

    1 4 3

  - 2 8

Likely Student Answer:

    1 4 3

  - 2 8

= 1 2 5

Explanation of Misconception: Students with misconceptions about subtracting whole numbers would come up with 125 as the answer to this problem. This error is both conceptual and execution. While the student knows that subtracting means taking away one value from another, the approach with this problem is fundamentally wrong. With execution, the process breaks down because of the "smaller-from-larger" approach by the student. Correct understanding of subtraction includes the notion that the columnar order (top to bottom) of the problem cannot be reversed or flipped. For some students, subtraction entails subtracting the smaller digit from the larger digit in each column regardless of the arrangement of numbers.

What Teachers Can Do: The use of models as hands-on activities (e.g., base 10 blocks) in teaching addition and subtraction is highly suggested for teachers especially in the early grades. It should be made clear that conceptual understanding of addition and subtraction concepts is a foundation on which arithmetic and other branches of mathematics depend. It is therefore imperative that efforts be expended on teaching them right. Teachers should make clear the relationship between addition and subtraction. For example, a teacher could explain that taking the subtrahend from the minuend gives the difference and conversely, when the difference is added to the subtrahend, the result is the minuend. This could offer mathematical understanding necessary for future solving of equations. It is also important to introduce vocabulary at this juncture. Alternative words (e.g., take away, minus, difference, etc) that may mean the same thing as subtracting should be defined and used interchangeably. Because regrouping is such an important concept in subtraction, teachers should make efforts in using models in illustrating it. Teachers should also emphasize that students check the results of problems they solve. This will enable them discover what they are doing wrong. For example, in the above problem, if the student had used addition procedure to check the answer, the process would have revealed that the solution and reasoning are incorrect. The student will discover that the addition of the difference (answer, 125) and the subtrahend (28) cannot produce the minuend (143) because $153 \neq 143$.

Research Note: Fuson et al (1997) did a meta-analysis based on some 4 projects involving multi-digit numbers in addition and subtraction and documented the following: (a) Multi-digit subtraction seems to be more difficult for children than multi-digit addition. Some difficulties seem to be inherent, and some may result from particular aspects of classroom activities, such as an emphasis on a take-away meaning. (b) Children may also incorrectly generalize attributes of addition methods to subtraction; this may be exacerbated if addition is experienced for a long time before subtraction. (c) One inherent source of difficulty in subtraction is the lack of commutativity of subtraction and the appearance of multi-digit numerals as constantly seductive single digits, especially in vertical form- this combination results in many children (and even adults, occasionally) subtracting the smaller from the larger number in a given column consistently or occasionally. (d) Experiences in two of the projects suggest that it may be better to intermix multi-digit addition and subtraction fairly early. For example, evidence from the PCMP project pointed to the fact that sustained experience solving problems requiring multi-digit addition before solving problems requiring multi-digit subtraction will for some children support an incorrect generalization of an addition solution method to subtraction. In the CBI project, such a separation of addition and subtraction problems might also have contributed at least somewhat to the CBI children's greater difficulty in devising a written method for subtraction than for addition.

Misconception 3

Addition of Fractions

Grades: 3 to 5

Question:

$$\frac{3}{7} + \frac{1}{2}$$

Likely Student Answer:

$$\frac{3}{7} + \frac{1}{2} = \frac{4}{9}$$

Explanation of Misconception: This misconception has to do with misapplication of rules to wrong situations. Students had probably learned in the past that when adding fractions that have the same denominator, the numerators could just be added while keeping the denominator the same. However, correct understanding of fractions with different denominators requires that the denominators of the fractions be adjusted to reflect same numerical values before the numerators are added. Another dimension related to this misconception is the fact that when students become familiar with certain algorithms involving whole numbers, the rules get in the way when students begin fractions.

What Teachers Can Do: To ensure conceptual understanding, teachers' initial strategy for teaching fractions should involve hands-on activities. Students should add fractions using manipulative materials. Fraction circles (pies) are excellent for modeling the addition of fractions. An addition problem with like denominators, 1/6 + 4/6, for example, can be demonstrated: *One pink piece (1/6) can be placed on the unit circle, followed by four pink pieces (4/6). The sum is represented by the fraction of the whole circle that is covered, which is 5/6.* To add fractions with unlike denominators, students should understand why both fractions are converted to those with like denominators. To demonstrate this idea, first model sums of fractions with unlike denominators using fractions that have equivalences that students already know. For example, model 1/2 + 1/4 by placing a yellow piece and blue piece on a unit circle. They can easily see that three-fourths of the circle is covered. To get students to move to common denominators, consider a task such as 5/8 + 2/4. Let students use pieces of pie to get the result of 1⅛ using any approach. Students will notice that the models for the two fractions make one whole and there is 1/8 extra. The key question to ask at this juncture is, "How can we change this problem into one that is just like the easy ones where the parts are the same?" For this example, it is relatively easy to see that fourths could be changed into eights. Have students use models to show the original problem and also the converted problem. The main idea is to see that 5/8 + 2/4 is exactly same problem as 5/8 + 4/8. Next, try some examples where both fractions need to be changed – for example, 3/7 + 1/2. Again, focus attention on rewriting the problem in a format similar to the ones in which parts of both fractions are the same.

Research Note: Hunker (1998) makes an excellent case for use of contextual problems and for letting students develop their own methods of computation with fractions. Hunker also stressed the need to allow students use other non-traditional ways to finds answers to simple addition problems. According to Hunker, "Students will continue to find ways to solve problems with fractions, and their informal approaches will contribute to the development of more standard methods." On their part, Bezuk & Cramer (1989) emphasized the need for students to estimate the size of the sum before they model the problem with manipulatives. They encouraged teachers to "implement more appropriate objectives for teaching fractions and to use instructional approaches that emphasize student involvement, including the use of manipulatives." (p. 160). They also emphasized the development of understanding before introduction to formal symbolic representations.

Misconception 4

Subtraction of Fractions

Grades: 3 to 5

Question:

$$\frac{3}{5} - \frac{1}{2}$$

Likely Student Answer:

$$\frac{3}{5} - \frac{1}{2} = \frac{2}{3}$$

Explanation of Misconception: This misconception is same as Misconception 3 above; this one being subtraction. The student performed the subtraction operation distinctly with the numerator producing 2 (3 -1 = 2) and the denominator producing 3 (5- 2 = 3). Again, the student applied the wrong algorithm in solving the problem and the lack of conceptual knowledge about fractions is responsible. The student could have manipulated the denominators to reflect same value before attempting to perform the subtraction operation. Other methods may also suffice.

What Teachers Can Do: In fractional problems, it is important to impress on students that the numerator tells the number of parts and the denominator the type of part. Premature attention to rules for computation should be discouraged. None of the rules helps students think about the operations and what they mean. Armed only with rules, students have no means of assessing their results to see if they make sense. Surface mastery of rules in the short term

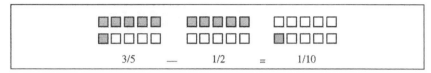

Figure 4.1. Illustration of 3/5 – 1/2 Using the Array Technique.

is quickly lost as the myriad of rules soon become meaningless when mixed together. The following strategies are suggested: (a) Begin with simple contextual tasks, (b) Connect the meaning of fraction computation with whole number computation, (c) Let estimation and informal methods play a big role in the development of strategies, and (d) Explore using models: Use a variety of models and have students defend their solutions using models. Teachers will find out that sometimes, it is possible to get answers with models that do not seem to help with pencil-and-pencil approaches.

The ideas gleaned from models will help students learn to think about the fraction and the operation, contribute to mental methods, and provide a useful background when they eventually get to the standard algorithm. One example of models would be the use of array of physical objects like chips to illustrate the concept of fractions with different denominators.

An array form could be constructed as shown in Figure 4.1. In the illustration, it shows that the fractions are first put in equivalent forms: 3/5 is 6/10 and 1/2 is 5/10. Thereafter, the subtraction operation is performed. The illustration shows that 3/5 – 1/2 = 6/10 – 5/10 = 1/10. Again, physically manipulating the objects (e.g., chips) to perform this task is beneficial to students.

Research Note: Cramer et al (2002) studied the effect of two different curricula on the initial learning of fractions by grades 4 and 5 students. One of the curricula is the commercial curriculum (CC) that could be described as traditional. The other is the Rational Number Project (RNP) curriculum that placed particular emphasis on the use of multiple physical models and translations within and between modes of representation – pictorial, manipulative, verbal, real-world, and symbolic. Students using RNP project materials had statistically higher mean scores on the posttest and retention test on four (of six) subscales: concepts, order, transfer, and estimation. The result also showed differences in the quality of students' thinking as they solved order and estimation tasks involving fractions. RNP students approached such tasks conceptually by building on their constructed mental images of fractions, whereas, CC students relied more often on standard, often rote, procedures when solving identical fraction tasks. The program of study by the CC students did not include a wide variety of materials or regular use of hands-on manipulative experiences but focused instead on pictorial and symbolic modes of representation. Also, there were substantial differences in the amounts of time devoted to various topics by teachers. For example, in the RNP group, a large amount of time was devoted to developing an understanding of the meaning of fraction symbol by making connections between the symbols and multiple physical models.

Misconception 5

Rounding Decimals

Grades: 4 to 6

Question: Round 525.25 to the nearest tenth

Likely Student Answer: 50

Explanation of Misconception: This student must have assumed that to correct to the nearest tenth means leaving the answer in two digits because tens are two digit numbers. The student therefore has made a structural error related to conceptual understanding. The student failed to realize that to round a decimal, one needs to first look at the digit to the right of the place one is rounding. If the number to the right is 5 or more, the number in that place is increased. The next thing to do is to drop the digits in all the decimal places beyond the place being rounded. For example, rounding 34.546 to the tenths place results in 34.5; rounding to the hundredths place results in 34.55; and rounding to the nearest tens give 30.

What Teachers Can Do: In teaching rounding of numbers, teachers should point out that a *th* separates *ten* from *tenth* (same with hundred and hundredth; thousand and thousandth). So, it is important for students to listen carefully to avoid confusing these names of number places. Too often, the process of rounding numbers is taught as an algorithm without reflection on why the algorithm makes sense. Children come to believe that to "round" a number means to do something to it or change it in some way. In reality, to *round* a number means that you substitute a "nice" number as an approximation for the more cumbersome original number. Teachers should remind students that when rounding to a specific value, they should avoid starting at the end of the number and round each place because this incorrect method may give a wrong answer. Remind students to underline the specific place value and consider only the digit to the right of the underlined place. In general, the teacher should emphasize the concept of place value.

Research Note: Steinle (2005) reported that LAB (Linear Arithmetic Blocks) are a good physical model to teach decimal numbers in which for example, 0.26 is made with 2 tenths and 6 hundredths (0.26 = 2 tenths + 6 hundredths). The report emphasizes the additive structure of the base ten numeration system. Steinle emphasized the need for students to be assisted with the concept of re-unitizing, for example 3 tenths equals 30 hundredths. Steinle also pointed out the need for constant use of the number line as it is possible to incorporate the various sets of numbers that has different focus with the curriculum, for example, whole numbers, common fractions, negative numbers and decimals. This longitudinal study found out that students who exhibit the misconceptions in decimals become more likely to persist with this behavior as they move to older grades and become less likely to move to expertise.

Misconception 6

Comparing Decimals

Grades: 4 to 6

Question: Which is greater? 3.215 or 3.7

Likely Student Answer: 3.215 is greater than 3.7.

Explanation of Misconception: Just as some students would indicate that 0.4123 is greater than 0.5, they would also likely state that 3.215 is greater than 3.7 because 3.215 has more digits than 3.7. Again, this misconception is related to structural error because the student lacks conceptual understanding – the student did not possess the knowledge related to number density. The student simply counted the actual number of digits and based their comparison on that fact. To the student, four numbers are more than two numbers, and hence the conclusion that 3.215 is greater than 3.7.

What Teachers Can Do: In teaching of decimals, teachers should emphasize the importance of lining up decimal points and comparing place value. They should start with decimals that have the same number of decimal place(s); for example, 2.5 and 2.6. The teacher should then progress to compare numbers like 7.4 and 7.04 which obviously would be more challenging and confusing than the first set of numbers (2.5 and 2.6). The teacher should point out that when decimals are lined up, for example in the case of 7.4 and 7.04, the 4 in the tenths place is greater than the 0 in the tenths place signifying that 7.4 is greater than 7.04. Next, the teacher should extend this approach and the understanding that students currently have to numbers with more than 2 decimal places. The importance of adding zeros to make the two numbers end in the same place value before comparing the numbers should be emphasized. In the above problem, students should write 3.7 as 3.700 before attempting to compare place values. The use of models such as the number line to represent values is also suggested. This will physically show students that when moving from zero to the right side of the number line, the number 3.215 comes before 3.7, and hence the latter is greater than the former. Also, the use of base 10 blocks and grids is suggested. The fundamental idea is for students to extend their place value concepts to decimals.

Research Note: A study by Steinle (2004) identified the "Larger-is-larger" behavior of students' misconception in which students chose the decimal with the most digits after the decimal point as the largest. Steinle suggested measurement context teaching of decimals, non-measurement context teaching of decimals, expanded notations, and avoidance of "staged" treatment of decimals. Steinle maintained that staged treatment of decimals, in which decimals with the same number of decimal places are considered, is unlikely to provide students with suitable experiences for them to appreciate number density. A useful non-measurement context to discuss decimals as Steinle pointed out is the *Dewey Decimal System* for locating books in a library (A student looking for a book numbered 510.316 will need to know that it will be found after 510.31 and before 510.32).

The study concluded by suggesting that teachers need to be aware that always rounding the result of a calculation to two decimal places can reinforce the belief that decimals form a discrete system and that there are no numbers between 4.31 and 4.32, for example. While it is often appropriate to round the results of a calculation to two decimal places, or to consider, for example, only two significant figures due to limitations of the initial measurements, students seem confused about when this is reasonable or appropriate. For example, to convert 3/8 to a decimal we can determine the result of the division of 3 by 8 and will need to rewrite 3 as 3.000 in order to find the result of 0.375. Hence, it is sometimes appropriate to consider 3 ones to be the same as 30 tenths, 300 hundredths, or 3000 thousandths (i.e., $3 = 3.0 = 3.00 = 3.000$). Yet on other occasions, this is inappropriate as the presence of the zeros indicates the precision with which a measurement is taken. Steinle suggests that these issues need to be discussed in the classroom.

Misconception 7

Multiplying Decimals

Grades: 4 to 6

Question: Multiply: 2.4 x 10

Likely Student Answer: 2.4 x 10 = 2.40

Explanation of Misconception: This misconception is associated with misapplication of rules and thus an arbitrary error related to conceptual understanding. The student probably had learned that multiplication by numbers having zeros (10, 100, 1000, etc.) involves moving the decimal point and placing zeros. However, this student does not know when it is appropriate to move or not move a decimal point. The student also does not know when and where to place zeros and hence the combining of the zero and the 4 to produce an answer of 2.40. The right knowledge of decimal multiplication requires the student to have the knowledge of how the position of the decimal point changes when multiplying by numbers having zeros. To simply move the decimal point one time to the right would have produced the right answer.

What Teachers Can Do: First, emphasize reasonableness of an answer by estimation. Teachers should have students come up with a reasonable estimate of a problem before attempting to solve the problem. After arriving at an answer, they should compare both their estimated value and the calculated value to see if there is any discrepancy. This will help students self-check and correct some errors they might encounter during computation. For example, it will be helpful if a teacher makes the student see that 2.40 is obviously much less that one of the multiplicands: (10). Second, because students would be confused as to where (direction) to move the decimal point when multiplying by a number with zeros, or the number of times to move, the teacher should apprise students that: (a) The multiplication of a decimal by 10 simply moves the decimal point one place to the right. Therefore, multiplication by 10n moves the decimal point n places to the right; and (b) The basic algorithm for decimals involves: — first, multiply the decimals as if they were whole numbers without regard to the decimal; second, determine the number of digits to the right of the decimal point in each of the decimals, and add these two numbers together; finally, the sum in the second step will be the number of digits to the right of the decimal point in the answer. Place the decimal point in the answer accordingly.

Research Note: While Irwin & Britt (2004) are skeptical of understanding of decimals by students because of the inappropriate generalizations that students make on the basis of prior knowledge, Moss & Case (1999) have suggested that starting students with percentages and moving from there to decimals yields positive dividends as documented by their Numeracy Project. In the project, students were discovered to be better equipped to operate with decimals using a variety of part-whole strategies. Lessons were focused on the place value of decimals, for example, using Unifix cubes in groups of 10 or pipes of different lengths. Steinle (2004) suggested the number slide as a powerful visual model for teaching division and multiplication by 10, 100, 1000, etc. It consists of a frame made from paper or cardboard. The names of the place value columns and the decimal points on the paper or cardboard are woven through the frame. Steinle says that the model is particularly helpful to kinesthetic and visual learners.

Misconception 8

More on Multiplying Decimals

Grades: 4 to 6

Question: Multiply: 0.1 x 0.1

Likely Student Answer: 0.1 x 0.1 = 0.1

Explanation of Misconception: Because 1 x 1 = 1, students are very likely to extend that fact to arrive at an answer of 0.1 for this problem and thus misapply mathematical rules. The conceptual understanding of decimals and decimal multiplication is lacking. The correct notion of multiplying two numbers that are each one decimal place would have been to obtain a product that has 2 places of decimals, and not 1 decimal place as displayed by this student.

What Teachers Can Do: Teachers should not attempt to explain this concept without some kind of models. Any hands-on activity that will illustrate the concept to the understanding of students is suggested. One such strategy is the use of grid paper (graph paper) to illustrate tenths and hundredths and also to perform multiplication of decimals. The teacher could either have students draw a 10 by 10 square grid or provide them with a ready-made grid paper. Tell students that the grid represent one whole. If one little square is colored in that outline, that is one hundredth or 0.01, two little squares is 0.02, and so on.

If ten of the little squares are colored, that would be the same as coloring in one column, or one row, we have a tenth, or 0.1. The teacher could use that model to show multiplication of decimals involving tenths. For example as shown in Figure 8.1, in the multiplication of 0.1 x 0.1, shade in one row in one color, say red. And then, shade in one column in another color, say blue. Where the two shadings intersect is the product of the two factors. Since 1 little square is shaded by both red and blue, we know that the product is 1

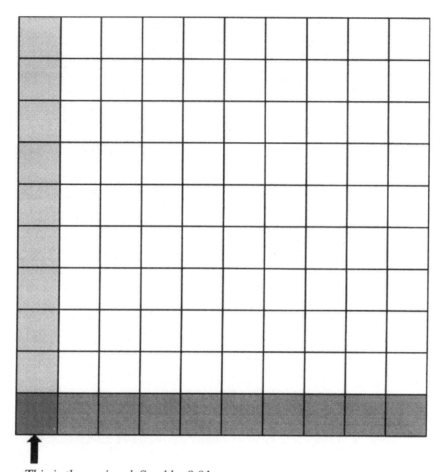

This is the region defined by 0.01

Figure 8.1. Multiplication Grid.

hundredth, or 0.01. Emphasize that this concept is true for multiplication of tenths only. Point out that it is not possible to represent and solve decimal problems that are either hundredths or thousandths with the 10 by 10 grid unless the values are approximated to the tenths value. Teachers could actually challenge students to explore this with more problems. Other varieties of grids could also be introduced to work problems that are not in the tenths category.

Research Note: Irwin (2001) investigated the role of students' everyday knowledge of decimals in supporting the development of their knowledge of decimals. Sixteen students, ages 11 to 12 were asked to work in pairs (one member of each pair a more able student and one a less able student) to solve problems that tapped common misconceptions about decimal fractions. Half the pairs worked on problems presented in familiar contexts and half worked on problems presented without context. A comparison of pretest and posttest revealed that students who worked on contextual problems made significantly more progress in their knowledge of decimals than did those who worked on noncontextual problems. In offering explanation to support the findings of the study, Irwin opined that problems presented in everyday settings provided the context needed for reflection on the scientific concept of decimal fractions. This reflection, for example, on the meaning of money given to more than two decimal places required students to reflect on how their existing concepts related to scientific knowledge. Such problems may have provided the reflection required for expanding their knowledge of decimal fraction. The findings suggest that complete understanding of decimal fractions requires multiplicative thinking, which is not natural but requires a reconceptualization of the relationship of numbers from that required in additive relationships. The study concluded by advising teachers to be aware of student's everyday knowledge and the misconceptions they may have on their way to achieving scientific knowledge of decimals. Teachers need to pose questions and mediate dialogue, interweaving everyday knowledge with scientific knowledge.

Misconception 9

Division of Decimals

Grades: 4 to 6

Question: Divide
        0.5 / 0.05

Likely Student Answer: 0.5 / 0.05 = 1

Explanation of Misconception: Children may have learned in the past that when a number is divided by itself, the answer is 1. This algorithmic fact is erroneously used by some students even when the numerator and the denominator are of different values as shown in this case. This student divided the 5 in the numerator by the 5 in the denominator to arrive at 1 as an answer without regard to the positions of the decimal points. The right knowledge of this concept requires that the decimals in both the numerator and the denominator be cleared by moving them equal number of times. This would have left us with 50/5 in the above problem.

What Teachers Can Do: Teachers should ensure that students have a good number sense with decimals and the values attached to tenths, hundredths, and thousandths. For example students should know that .5 is 50% and thus 1/2. Also, .05 is 5% and thus 1/20. To develop this type of familiarity, students do not need new concepts or skills. They do need the opportunity to apply and discuss the related concepts of fraction, place value, and decimals in activities. Teachers should give students the opportunity to illustrate the quantities on grid, on number line, and even circle graph.

   For example, while students will see that .5 covers 50% or 1/2 of a 10 x 10 grid, they will also practically see that .05 translates to only 5 pieces of squares being covered on the same grid. Other than use grids to help students understand what the values of 0.05 and 0.5 represent, the teacher could also

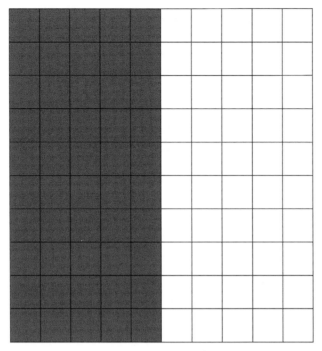

Figure 9.1a. Pictorial Representation of 0.5.

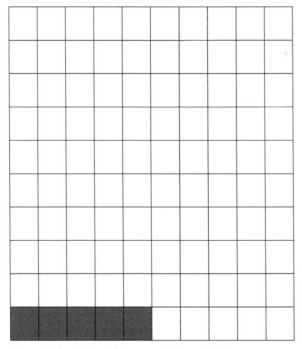

Figure 9.1b. Pictorial Representation of 0.05.

use a number line or a circle graph to illustrate. After ensuring that students understand what these values represent, the next thing is to ensure that they understand what the question wants them to do. The teacher should emphasize that $0.5 \div 0.05$ basically means that one uses 0.05 as the unit to measure 0.5. The technique of moving decimals should also be taught. A reasonable algorithm for division is parallel to that for multiplication: *Ignore the decimal points, and do the computation as if all numbers were whole numbers. When finished, place the decimal by estimation.* If students have a method for dividing by 5, they can divide by .5, .05, and even .55. It is therefore important to teach divisibility of numbers, especially from 2 to 9.

Research Note: Contextualized teaching has been reported on several times as indeed necessary for teaching decimals. However, the literature points out that children of different socio- economic levels require different context problems. Resnick, Bill, Lesgold, & Leer (1991) suggested that children from minority cultures are less likely than students from a dominant culture to spontaneously use the knowledge they have learned outside school when learning new concepts in school. As reported by Irwin (2001), teachers of low income or diverse classroom need to be aware of students' everyday knowledge and any misconceptions developed on the way to achieving scientific knowledge of decimal fractions. Irwin states that there is need to recognize that what amounts to everyday knowledge for one group may not be everyday knowledge for another group: baseball enthusiasts know baseball statistics and basketball enthusiasts know basketball statistics. So students will not draw on their everyday knowledge if it does not seem necessary to do so. For example, Britt, Irwin, Ellis, & Richie (1993) found that students from lower economic areas had more difficulty than did students from more affluent areas in understanding decimal fractions. In that study, not only did the students attending a school in lower income area have poorer understanding at the start of the school year, but they also made less progress during the year: 22% to 32% compared to middle income areas which was 62% to 93% of students that understood the concept of hundredth or more complex decimal relationships.

Findings by Brekke (1996) were more contradictory when emphasizing teaching of decimals within context, e.g., money. The researcher stated that teachers regularly claim that pupils manage to solve arithmetic problems involving decimals correctly if money is introduced as a context to such problems. The study pointed out that it is doubtful whether continued reference to money will be helpful when it comes to developing understanding of decimal numbers; on the contrary, this can be a hindrance to the developments of a robust decimal concept. [Note: The author does not necessarily agree with the findings of Brekke].

Misconception 10

Percent Problems

Grades: 4 to 6

Question: Convert 0.225 to percent

Likely Student Answer: 0.225 = 0.00225%

Explanation of Misconception: The student probably had learned in the past that converting from decimal to percent involves moving the decimal point. This student therefore applied a known rule in the wrong fashion. While the student is right about moving the decimal point two places, there is confusion as to what direction to move. The student should have moved the decimal point to the right and arrive at 22.5% as the correct answer. This misconception and others such as calculating percent discounts (arriving at $60 for an article on sale 20% off if the original price is $80) are quite common. In this kind of discount problem, the student may just simply subtract 20 from the original price to arrive at $60 as an answer.

What Teachers Can Do: Teachers are encouraged to relate percent problems to real-world contexts and should constantly make references to percents as they appear in stores, newspapers, and television. In addition to real-life problems, it is important for teachers to follow the following maxims for a unit on percents: (a) Limit the percents to familiar fractions (halves, thirds, fourths, fifths, and eighths) or easy percents (1/10, 1/100), and use numbers compatible with these fractions. The focus of these exercises is the relationships involved, not complex computational skills, (b) Do not suggest any rules or procedures for different types of problems. Do not categorize or label problem types, (c) Use the terms: *part*, *whole*, and *percent* (or *fractions*). Help students see these percent exercises as the same types of exercises they did with simple fractions, (d) Require students to use models and drawings to explain their solutions. It is better to assign a couple of problems that require drawing and explaining rather than give many problems requiring only computation and answers. Remember that the purpose is the exploration of relationships, not computational skills, and (e) Encourage mental computation.

Furthermore, establishing that decimals, fractions, and percents are just different ways of showing the same value is important. The teacher could provide a table of commonly occurring decimals to students so they can see a pattern. For example: 1% = 0.01 = 1/100, 5% = .05 = 1/20, 10% = 0.1 = 1/10, and so on. Again, having different models like the grid, pie graph, and line graph to actually represent each of these quantities is important. Remind students that percent means *per 100*, so a percent is a ratio comparing a number to 100. For example, if a problem asks students to find how much they are saving for a backpack of original price of $50 but is on sale for 30% off, the use of grid as a model is ideal.

Have students draw a 1- by-10 rectangle on a grid paper as shown in Figure 10.1. Label the units on the right from 0% to 100%. Since $50 represents the original price, mark equal units from $0 to $50 on the left side of the rectangle. Draw a line from 30% on the right side to the left side of the model and shade the portion of the rectangle above the line. The model shows that 30% of $50 is $15. This is suggested for teachers as a first step of working with percents before embarking on specific algorithms and procedures. The teacher should point out to students that if they plan to use a decimal model to find a percent of a number that is not divisible by 10 or 5 as done in this example, they may need to modify the intervals on their decimal model or use their decimal model solely for estimation purposes.

Research Note: Researchers emphasize "Big Idea" as of paramount importance when teaching percent concepts, a phrase popularized by Carnine (1994). In an instructional program, "Big Ideas" are the important concepts that provide the anchors and connections for the to-be-learned information (Baker, et al., 1994). According to Carnine, teaching the "Big Idea" that underlies numerous, but related instructional objectives, should replace teaching for coverage of a broad curriculum area (where instructional objectives are frequently taught as a series of unrelated skills) with teaching a few interconnected "Big Ideas" which will be taught thoroughly. By structuring instruction regarding fractions and percents around the "Big Idea" of division, Carnine states that teachers can provide clearer explanations of percents (e.g., by demonstrating that 3/4 is a fraction that represents 3 parts of a whole which was originally divided into 4 equal parts and represents 75% of the whole). Carnine states that by stressing the underlying unifying concept of division, students learn that there are explicit links between fraction and decimals and percents, and the mathematical curriculum objectives that are typically taught

Percent Problems

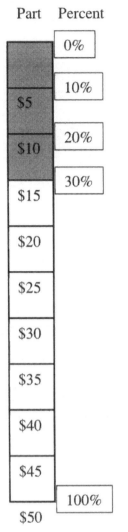

Figure 10.1. Part–Whole Relationship.

as discrete, unconnected skills, can, and should be connected to mathematical skills already mastered. Not only does this particular "Big Idea" make sense, but practical aspects of instruction (e.g., manipulation and computation of such problems with calculators) are enhanced.

Misconception 11

Division by a Fraction

Grades: 6 to 8

Question:

$$-9 \div \frac{1}{3}$$

Likely Student Answer:

$$-9 \div \frac{1}{3}$$
$$= -9 \times \frac{1}{3}$$
$$= -3$$

Explanation of Misconception: In this problem, the student erroneously multiplied -9 by the divisor instead of the reciprocal of the divisor. These kinds of problems are solved using the popular "Keep, Change, and Flip" mnemonic in which the dividend is kept the way it is, the division sign is changed to multiplication, and the divisor is flipped. Obviously, the student remembered to "keep" and "change" but did not "flip" the 1/3 to 3/1, and hence the wrong answer.

What Teachers Can Do: The teacher should emphasize that multiplication is the inverse of division and vice versa. Before exposing students to the mnemonic idea, teachers should insist on conceptual understanding of dividing a negative number by a fraction. Since division is the inverse operation of multiplication, every division problem can be written as multiplication by a reciprocal: $x \div y = x \cdot (1/y)$. Therefore, the rules for multiplication of integers and of division are similar: (a) When integers with the same sign are divided, the result is a positive number, and (b) When integers with different signs are divided, the result is a negative number. Teachers should help students understand why dividing is the same as multiplying by the reciprocal through models and examples: *Use three circular pieces of paper to model three large cookies or pizzas. If each portion is to be half a circle, how many portions are there?* Help students connect this to the representation of $3 \div 1/2$. Write the following expression on the board: $8 \div 2$. Make students agree that the quotient is 4. Then, rewrite the 8 and the 2 as fractions: $8/1 \div 2/1$. Compare this expression with the expression $8/1 \cdot 1/2$, which has product of 4.

Research Note: Researchers have identified lack of connection to students' lives and situations in which they could use them as a factor affecting the learning of division by fractions. Van de Walle (2004) emphasized the use of problem solving as a strategy. He specifically suggested that story problems be used in teaching division by fractions. Ma (1999) identified the "invert the divisor and multiply" concept as one of the most poorly understood procedures in K-8 curriculum. In presenting the result of the study on teaching this concept by Chinese and U.S. teachers, Ma showed that Chinese teachers do not only uses the algorithm, but they also understand why it works. On their part, U.S. teachers were found to be sadly lacking in their understanding of fraction division and could not display how the concept works.

Behr, Khoury, Harel, Post, & Lesh (1997) researched on pre-service teachers' understanding of the operator construct of a rational number: finding 3/4 of a pile of 8 bundles of 4 counting sticks using two different subconstructs. One is the showing of 3/4 of the number of bundles (duplicator/partition-reducer [DPR] subconstruct) and the other is the showing of 3/4 of the size of each bundle (stretcher/shrinker [SS] subconstruct). This study provides confirming instances that candidates uses these two rational number operator subconstructs. The SS strategies are identified when the rational number, as an operator, is distributed over a uniting operation. With these SS strategies, rational number is conceptualized as a rate. The result of the study also showed that the SS strategies were used less often than the DPR strategies.

Misconception 12

Ordering Fractions

Grades: 3 to 6

Question: Arrange in order from greatest to least: 1/3, 1/2, 1/6.

Likely Student Answer: 1/6, 1/3, 1/2.

Explanation of Misconception: The student is applying the knowledge of natural counting numbers and number density to fraction density. To the student, 6 is greater than 3, and 3 is greater than 2, hence 1/6 > 1/3 > 1/2. This misconception is common in grades 4 and 5 where students are introduced to fractions. One reason for the misconception is the fact that the fractions (1/3, 1/2, 1/6) are not the same as the ones they use to know and compare. For example, they must have been familiar with same-denominator fractions in which they just compare the numerator values (e.g., 2/6, 1/6, 3/6, etc). With different denominators, the right conceptual understanding requires the knowledge of transitive strategy in ordering fractions. The process involves comparing the fractions to external values by reflecting that fractions are one "piece" away from the whole unit.

What Teachers Can Do: The difficulty with fractions should not be surprising considering the complexity of the processes involved. Children must adopt new rules for fractions that often conflict with well established ideas about whole numbers. For example, with whole numbers, 3 is greater than 2; but with fractions, students are again meant to construe that 1/3 is less than 1/2. When comparing fractions of this type, children need to coordinate the inverse relationship between the size of the denominator and the size of the fraction. The teacher should use models to, for example, make students realize that if a pie is divided into three equal parts, each piece will be smaller than when a pie of the same size is divided into two equal parts. With fractions, the *more* pieces, the *smaller* the size of each piece. As pointed out earlier, one difficulty is that the techniques for ordering fractions with like denominators (e.g., 1/6, 2/6, and 5/6) and to fractions with unlike denominators (e.g., 1/5, 2/7, and 5/8) are quite different. For example, a student might reason that the fraction 5/7 is greater than 2/7 because 5 is greater than 2. However, that understanding cannot be extended to comparing fractions such as 5/7 and 2/9. Therefore, to think quantitatively about fractions, students should know something about the relative size of fractions. They should be able to order fractions with like and unlike denominators as well as to judge if a fraction is greater than or less than say, 1/2. They should know the equivalents of 1/2 and other familiar fractions. The acquisition of a quantitative understanding of fractions should be based on students' experiences with physical models (e.g., fraction strips) and on instruction that emphasizes meaning rather than procedures.

Research Note: Bezuk & Cramer (1989) emphasized the need for informal ordering schemes to estimate fractions quickly and to judge the reasonableness of answers. According to the researchers, many pairs of fractions can be compared without using formal algorithm such as finding common denominator or changing each fraction to decimal. Behr, et al (1983) recommended using a variety of manipulatives to develop fractions concepts. According to the researchers, the use of more than one manipulative enhances students' understanding and promotes abstraction of the concept from irrelevant perceptual features of the manipulative, such as color, size, or shape.

Misconception 13

Least Common Multiple (LCM)

Grades: 3 to 8

Question: What is the LCM of 1/4 and 1/10?

Likely Student Answer: LCM of 1/4 and 1/10 is 2

Explanation of Misconception: The misconception here has to do with concentrating on the "least" in the problem. Students often think that they need to find the least number that can divide both denominators. They usually do not realize that the Least Common Denominator (LCD) of two or more fractions is the Least Common Multiple (LCM) of the denominators.

What Teachers Can Do: The use of non-traditional approaches is suggested for teaching the concept. In the above problem, we know that 2 is an answer to the Greatest Common Factor (GCF) of 4 and 10 (which is a related concept). The teacher could use Venn diagrams as a means of distinguishing between the LCM and GCF.

Having introduced students to the concept of prime factorization, the teacher could extend that knowledge to a two-Venn system in finding the LCD and GCF of numbers. As shown in Figure 13.1, there is a union set: {2 x 2 x 5} and an intersection set: {2}. The teacher should explain what they represent. For example, saying something to the effect that the number 2 that is an intersection (middle) of 4 and 10 in the Venn diagram is the Greatest Common Factor (GCF). On the other hand, all the prime factors appearing in both Venns including the intersection represents the Least Common Multiple (LCM) of the numbers. This is {2 x 2 x 5} and equal to 20. A teacher could extend the concept to situations where there is no intersection between the prime factors of two numbers in a Venn diagram, say, 4 and 21. Because this is another source of misconception, the teacher should hint that there is no

Figure 13.1. Venn Diagram Showing Prime Factors of 4 and 10.

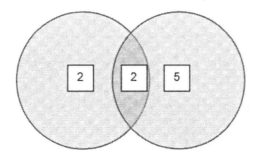

intersection of sets in the Venn diagram for the numbers (4 and 21) and hence the GCF is 1 but not zero as most students would erroneously deduce.

Research Note: Brown et al (2002) interviewed prospective elementary school teachers with questions on number theory and found that three approaches were used to find the least common multiple between two numbers: (a) set intersection – creating an ordered list of consecutive multiples for each number and finding the first one that appears on both lists; (b) creating a multiple and divide – creating an ordered list of consecutive multiples of one number and simultaneously checking each new entry for divisibility by the second number; (c) prime factorization – comparing the prime factorization of the two numbers and finding the minimal product of prime powers that contains both of their factorizations (often referred to as "the higher exponent" rule). Brown (2002) remarks that the first two methods can be wieldy because it may require a large number of steps, but that, on the other hand, justifying these two first methods is much more intuitively done than explaining why the "higher exponent" rule work.

Brown et al (2002) concluded that to be able to use multiplicative structure in reasoning, one needs to deal flexibly with prime factorizations and be able to identify multiples and factors through the application of meaningful process. According to the researchers, "A major goal of experiences of this type would be not only to teach students procedures (actions) needed for dealing with numbers in primed-factored form, but to get students to reflect on these actions sufficiently and in productive enough ways that they are able to move beyond successful actions to processes and objects." (p. 78). Researchers are pointing this out because they found that students often change numbers from prime-factored representation to base-ten representation in order to perform mental operations on them. That is, they were not able to make connections between actions, processes, and objects as a *schema* – a process termed "thematization." (Ferrari, 2002).

Misconception 14

Addition of Decimal Numbers

Grades: 3 to 6

Question: Add: 0.77 + 9 = ?

Likely Student Answer:

```
   0.77
+     9
   ────
   0.86
```

Explanation of Misconception: When decimal points are not lined up for students, they tend to add or subtract decimals the wrong way. This is an execution error because the student has the knowledge related to the addition algorithm but the process of adding a decimal number and a whole number breaks down when attempted. The error in rearranging into a vertical format by the student is the source of the problem. The right conceptual knowledge requires that the student line up decimals appropriately by rewriting: 0.77 + 9.00.

What Teachers Can Do: This concept should be taught with models, e.g., the number line. The number line could be valuable in helping students understand the concept of addition of decimal numbers. The first step for this kind of problem is to have students indicate the various points where .77 and 9 are on the number line and compare the solution (0.86) to see if it is reasonable. With the understanding that both addends ought to produce an answer greater than each addend, students will discover that 0.86 is not a solution of 0.77 + 9. Next, the teacher could have each student draw two different number lines graduated from zero to say, 20. After marking out .77 and 9 on the lines, they should impose one line on the other and make a lateral movement of one line's zero point to coincide with .77 of the second line by gliding. Ask students to explain what they observe. They will practically see that the length 0.77 added to the length 9 results in total length of 9.77 on the number line. This will further reinforce the importance of emphasizing reasonableness of answers: 0.86 is in sharp contrast with a number that is approximately 10. As always, story problems are helpful with decimal problems. After giving students several opportunities to solve addition story problems, the following activity may be used in conjunction: Give students a sum involving different numbers of decimals places, say, 45.23 + 5.5 + 0.213. The first task is to make an estimate and explain the way the estimate is made. Next, ask students to compute the exact answer and explain how that was done without calculating aids. Finally, students should devise a method for adding decimal numbers that they can use with any two numbers. When students have completed these tasks, the teacher should have them share their strategies for computation and test them on a new computation problem provided by the teacher.

Research Note: Teaching decimals in a comprehensive way should give students the opportunity to flexibly utilize and link a variety of representations concerning decimal numbers: linear model such as number line, manipulatives, symbols, and currency (Thomson & Walker, 1996). Even though research findings often disagree with regard to the importance of number line as a didactical model in general and as means of representing integers and rational numbers, (eg., Raftopoulos, 2002), the knowledge of how to use the number line as a means of representing the decimal is quite necessary since it contributes to the development of concepts not only related to the identification and comparison of decimals but to the ability to perform operations as well (e.g., Thomson & Walker, 1996).

A study by Michaelidou et al (2004) confirmed that students are faced with difficulties in addition of decimal numbers. The result of the study indicates that addition tasks and especially those that include translation procedures presupposed the ability of performing the operation of addition, the ability to represent the adders and the operation on a number line and the ability to manipulate and interpret the information represented by the number line and its features. Consequently, addition tasks might have caused a great cognitive load to the students dealing with them. The study showed no statistically significant differences between the means of students that solved the problem with restriction to the solution procedure and the students that were asked to use the number line in order to solve the same decimal addition problems. They concluded that the fact that there is no significant difference between the two groups might equally indicate that the number line must have caused more difficulties to some of the students in the experimental group. According to the researchers, it is highly likely that the necessity to manipulate and interpret the number line and its features may have added to the cognitive load of students and led them to erroneous approaches concerning the problems.

Misconception 15

Subtraction of Integers

Grades: 5 to 8

Question: -12 − (-3) = ?

Likely Student Answer: -15

Explanation of Misconception: The correct knowledge of this concept requires that students subtract -3 from -12 to obtain -9. Instead, the student used the two negative signs to produce a positive sign and therefore added the 12 and the 3 to obtain 15. Thereafter, the negative sign is moved to the 15 and thus the answer of -15. This is a misconception necessitated by misapplication of rules. The conceptual understanding of operations with integers is lacking.

What Teachers Can Do: Teach the concepts of addition and subtraction that involve negative numbers with models; e.g., Red and Blue counters representing negative and positive numbers respectively. A teacher could use this scenario: *"If 12 red counters represent -12 and you remove 3 counters, that means subtracting -3 from -12. The result is 9 red counters represented by -9."* This approach may also be helpful when the question is of the form: 12-(-3). Students will be more challenged with this type of problem because it involves subtracting a negative number from a positive number. In this case, the use of counters is also suggested. The students should be introduced to the concept of *zero pairs*. Because there are no three negative counters to subtract from 12 positive counters, zero pairs should be created to add on. By creating three zero pairs, we now have -3 (three negative counters) to subtract. Thus we are left with 12 positives and 3 positives that add up to 15 as shown in Figure 15.1.

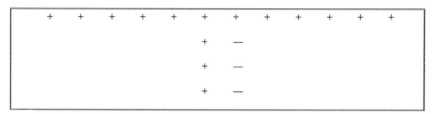

Figure 15.1. Diagrammatic Representation of 12 - (-3).

The diagrammatic representation is important. With it, students can see how the zero pairs are added and how the negative numbers are subtracted. Another model that teachers could use is the number line. With teaching and learning of integer operations, the use of the number line involves rules: where to start counting, direction to move, and the distance to move. To make the process less burdensome for students, explicitly stated rules that are handy (put them on poster card and hang it on the wall) is always a good idea. Teachers should also allow students to use the method with which they are most comfortable: counters, number line, analytically, etc. Teachers should not emphasize right answers but should always demand explanations from their students.

Research Note: Van de Walle (2004) pointed out that although the two models of teaching integer addition – models and number line appear different, they are alike mathematically. Integers involve two concepts of quantity and opposite: *Quantity* is modeled by the number of counters and *opposite* is represented as different colors or different directions. He stated that many teachers who have tried both models with their students report that students find one model easier and the other harder. As a result, they decide to use only the model that students like or understand better. Van de Walle remarked that this should not be the case. It may be that students find the operations easier when they use counters. This is not the same thing as understanding integer operations. Van de Walle suggests that students should experiment with both models and, perhaps even more important, discuss how the two are alike. A parallel development using both models at the same time may be the most conceptual approach. A problem-solving approach for integers was also emphasized. This according to Van de Walle can be done by assigning, say, half the class to one problem and to use models; the other half of class assigned to the same problem but uses the number line. When solutions have been reached, the groups can compare and justify their results. According to Van de Walle, incorrect ideas will surface and the learning that comes from the discussion and clarification are beneficial.

Misconception 16

Converting Linear Units

Grades: 2 to 4

Question: Convert 40 cm to meters

Likely Student Answer: 40 cm = 4 meters

Explanation of Misconception: Students should be acquainted with the customary units of length (inches, feet, yards, and miles) and the metric units of length (millimeters, centimeters, meters, and kilometers). Obviously, the answer by this student shows the contrary. The student misses the idea that 100 centimeters is equal to 1 meter and dividing the 40 cm by 100 would have produced the right answer. This is an arbitrary error associated with lack of conceptual knowledge.

What Teachers Can Do: Perhaps the biggest error in measurement instruction is the failure to recognize and separate two types of objectives: First, understanding the meaning and technique of measuring a particular attribute, and second, learning about the standard units commonly used to measure that attribute. These two objectives can be developed separately; when both objectives are attempted together, confusion is likely. It is important to expose students to nonstandard units because informal units provide a good opportunity for understanding standard units. Knowledge of standard units is a valid objective of a measurement program and must be addressed. Students must not only develop familiarity with standard units but must also learn appropriate relationship between them. Helping students develop familiarity with the most frequently used units of measure should be an important goal of instruction. Other goals includes: (1) Acquainting students with basic idea of size of commonly used units and what they measure, (2) Knowing what is a reasonable unit of measure in a given situation; and (3) Keeping emphasis

	Metric System	**Customary System**
Length	Millimeter, Centimeter, Meter, Kilometer	Inch, Foot, Yard, Mile
Area	Square Centimeter, Square Meter	Square Inch, Square Foot, Square Yard
Volume	Cubic Centimeter, Cubic Meter	Cubic Inch, Cubic Foot, Cubic Yard
Capacity	Millimeter, Liter	Ounce, Cup, Quart, Gallon
Weight	Gram, Kilogram, Metric Ton	Ounce, Pound, Ton

Figure 16.1. Commonly Encountered Units of Measurement.

on those relationships that are commonly used, such as inches, feet, and yards or milliliters and liters. Teachers should also emphasize conversion between units (For example: 12 inches = 1 feet, 3 feet = 1 yard, 5280 feet = 1 mile, 1 mile = 1760 yards, 1 meter = 39.37 inches, etc).

Activities do help students develop familiarity with standard units. A teacher could give students a model of a standard unit and have them search for things that measure about the same as that one unit. For example, to develop familiarity with the meter, give students a piece of rope 1 meter long. Have them make lists of things that are about 1 meter. Keep separate lists for things that are a little less (or more) or twice as long (or half as long). Encourage students to find familiar items in their surroundings. In the case of lengths, be sure to include circular lengths. Later, students can try to predict if a given object is more than, less than, or close to say, 1 meter. The same activity can be done with other unit lengths. The teacher should enlist the help of parents to assist students in finding familiar distances that are about the distance of a reference unit. For capacity units such as cup, quart, and liter, students need a container that holds or has a marking for a single unit. They should then find other containers at home and at school that hold about as much, more, and less.

Research Note: Joram et al (2005) reported on third grade students' use of two different strategies for measurement estimation. The researchers instructed one group of students on the use of the reference point strategy and a second group received instruction on the guess-and-check procedure. The main focus of the reference point instruction was to help students develop more accurate representations of linear measurement units, to show how to correctly iterate a reference point estimate, and to use reference points flexibly. The guess-check group completed many of the same activities as the reference point. However, the focus for these students was on the use of standard instruments for measuring and estimating, the correct measure for measuring, and on estimating accurately by receiving feedback on estimate. The researchers found out that the use of the reference point strategy was statistically associated with greater estimation accuracy both in the interview and group test settings. In addition, students who estimated using reference points had more accurate representations of standard linear units. Their findings suggest that, relative to students who do not use reference points estimate, those students who use the reference point strategy have enhanced representations of linear units which leads to more accurate estimates. When standards units (or multiples of standards units) are represented by reference points, they seem to be more easily recalled and imagined than their corresponding standards units. The researchers concluded that having a "feel" for standard linear units, represented by familiar objects, may form a foundation for two-dimensional measurement concepts (area), and three- dimensional measurement concepts (volume).

Misconception 17

Power to a Base

Grades: 5 to 8

Question: Simplify:
$$3^4 = ?$$

Likely Student Answer: $3^4 = 3 \times 4 = 12$

Explanation of Misconception: This misconception stems from structural errors associated with lack of conceptual knowledge. The student thinks that power of an exponent acts as a multiplier (that is, multiply the base and the exponent). The right conceptual thinking require that the power of exponent tells of how many times the base is used as a factor.

What Teachers Can Do: Point out to students that exponents are powers to which a base is raised and two or more numbers that are multiplied together form factors. When the same factor is used, one may use an exponent to simplify the notation. Also, point out that the representation of 3 x 4 which can be explained as 3 groups of 4 or 4 groups of 3 has a sharp contrast with 3^4. Tell students that one way to represent 3^4 as "groups of" would be to first of all list the factors: 3 x 3 x 3 x 3 and we would therefore have 3 groups of 3 multiplied by 3 groups of 3 to arrive at 81. Showing a pattern would also make students confirm that 3^4 cannot equal 12. For example, the teacher could show that: $3^1 = 3$, $3^2 = 3 \times 3 = 9$, $3^3 = 3 \times 3 \times 3 = 27$, and $3^4 = 3 \times 3 \times 3 \times 3 = 81$. This illustration should make clear that while multiplication is "groups of," and thus repeated addition, exponents indicate repeated multiplication. One of the problems that teachers deal with is transitioning from basic arithmetic process of multiplication to algebraic processes of powers, factors, and exponents. Teachers should be cognizant of this salient transition in their instruction. For example, 3 x 4 can also be written as m x n and simplified when m = 3 and n = 4; and 3^4 can be written as m^n and could be assigned appropriate values for simplification. These kinds of representations help students in making transition from arithmetic to algebra. Because these ideas seem abstract enough, relating them to student's lives would be beneficial. For example, remind students that they probably used metric measurements before. Grams and kilograms and other metric units are related to each other by powers of 10 (e.g., one kilogram is 10^3 grams).

Scaffolding questions can also make the procedure clearer to students as they give direction and momentum to a lesson, clarify its purpose and keep students on task. For example, give each pair of student ten unit cubes (1x1x1). Have students use the units cubes to make squares and ask them: How many unit cubes are in the a square with side of 1?, 1; sides of 2?, 4; sides of 3?, 9. Thereafter, ask students to predict how many cubes would be in a square with 5 sides. The students should be able to establish a relationship between the number of unit cubes needed to build a square and the number of units per side: The number of squares per side multiplied by the same number. These kinds of activities coupled with effective scaffolding questioning could enhance understanding of exponents.

Research Note: Martinez (1988) addressed helping students understand factors and terms in algebra and offered strategies for errors and confusion by students. Even though students are introduced to the concepts of factors in elementary arithmetic, they are often confused when applying this understanding to algebra. Martinez suggests the approach that builds directly on students' knowledge of arithmetic; a three- phase pedagogical strategy that promotes clear differentiation: (1) exploring the concept, (2) inventing mental structures to deal with the concept, and (3) discovering application of the concept in a variety of ways. Also, verbalizing mental processes can help facilitate the transfer from arithmetic to algebra applications. For example, students begin by explaining what may seem obvious to them – what they must do to solve the arithmetic exercise, then they consciously develop the parallels to solve the algebra exercise.

Martinez identified the following guidelines for the development of exercises that address errors that students are making: (a) work from the familiar to the unfamiliar and from simple to complex, (b) focus on one idea at a time, using that idea as a constant throughout the set or series of sets; (c) use repetition to emphasize relationships and to highlight differences; (d) increase complexity by easy steps to build pattern; (e) write several sets of exercise for key concepts to demonstrate different applications and perspectives; and (f) ask students to verbalize their mental processes, either aloud or in paper, to ensure understanding. According to Martinez, using the explore-invent-discover strategy and emphasizing relationships between what students learned in arithmetic and what they are learning in algebra reduces many students' confusion about factors. Emphasizing the importance of verbalizing, Martinez states that the act of verbalizing or translating operations from symbols to English and back to symbols is helpful. Not only does translation lend practice in critical English skills, but it also helps students understand that a concept can be represented in a variety of ways.

Misconception 18

Order of Operations I

Grades: 5 to 8

Question: Simplify the expression
$$30 \div 6 - 1$$

Likely Student Answer:

$$30 \div 6 - 1$$
$$= 30 \div 5$$
$$= 6$$

Explanation of Misconception: The rule for order of operation of numbers was not followed by this student. The student subtracted 1 from 6 before diving instead of dividing before subtracting. This is an arbitrary error of conceptual knowledge as the student veered from the conventions necessary to solve the problem.

What Teachers Can Do: The basic approach to this concept is to follow the rules: (1) Evaluate the expressions inside grouping symbols, (2) Evaluate all powers, (3) Multiply and divide on order from left to right; and (4) Add and subtract in order from left to right. In addition to these rules which by the way are easy for students to forget, teachers should resort to scaffolding questions to make the concept clear and encourage students to justify the steps in their solutions. With scaffolding questions, teachers can redirect students' thinking. The teacher could write the question on the board and ask: What is the value of expression if the division is done first? Next, the teacher should ask: What is the value of the expression if the subtraction is done first? A further question could be: How do you know which operation to perform first? This should be the caveat to start talking about the order of operations. The teacher should then point out the correct order and have students work similar problems. The teacher should model justifying steps taken to solve a problem. That way, students with misunderstanding of the concept will see that trying to justify an incorrect step could reveal what they needed to do right. For example, if a student decides to subtract first in the problem above, the student may discover through justifying steps that it is not only counter intuitive but does not follow the conventional rules of operation. The student can as a result, adjust the steps and solution. The teacher should provide assistance only when it is necessary to do so. Another strategy is to give students the opportunity to construct their own questions. Not only does the practice activate the schema necessary for the concept, but it also helps students reflect when working problems and can therefore self- check their solution paths.

Research Note: Carpenter et al (2003) identified two problems with order of operations. First is that even after students have been exposed to conventions of the order of operation, many students fall back on calculating from left to right regardless of the operations and order required of the question. Another problem identified by the researchers is that students can be become too rigid in applying the rules once they have learned them. They suggested holding on the order of operations until the associative and distributive properties are introduced. One of the reasons that it may be productive to wait until children use the associative and distributive properties before talking about order of operations is to provide students alternatives to calculating in a specific sequence without thinking. The order-of-operation conventions apply in cases where it makes a difference what order the calculations are done in. But there are many cases in which it is possible and perhaps easier to follow a different sequence of calculations. They gave an example of: 35 x 24 ÷ 12. According to them, a slightly different sequence can be followed because the answer is the same as it would be if the calculations were carried out following the conventions. For example, it is easier to divide 24 by 12 before multiplying, and the answer is the same as multiplying and dividing from left to right.

Carpenter (1989) also pressed for the importance of problem solving. According to Carpenter, teaching the concept through problem solving can encourage students to refine and build into their own processes over a period of time, as their experiences allow them to discard some ideas and become aware of further possibilities. As well as developing knowledge, the students will also develop an understanding of when it is appropriate to use particular strategies.

Misconception 19

Order of Operations II

Grades: 5 to 8

Question: Simplify the expression
   $12 \div 3 \times 2$

Likely Student Answer:

   $12 \div 3 \times 2$

   $= 12 \div 6$

   $= 2$

Explanation of Misconception: The student performed the multiplication operation first instead of the division which the order of operation requires. The popular mnemonic device (PEMDAS = Please Excuse My Dear Aunt Sally) which has the multiplication first must have prompted the solution route of this student. But rather than PEMDAS, it is the Left-to-Right rule that applies to this problem because only multiplication and division that has same priority are present.

What Teachers Can Do: In addition to the strategies discussed in Misconception 18, it is important to involve vocabulary. Possibly put out a word wall for students to familiarize themselves with various terms. Discuss with students the importance of consistency in mathematics. For example, $12 \div 3 \times 2$ must always have the same value 8 when simplified correctly. The order of operations ensures that the value of an expression is always the same, regardless of who calculates the value. By way of emphasizing the point, teachers could ask students if they can think of any rules similar to the order of operation that affects their lives. Driving rules are an example. Ask students to consider what driving would be like if all drivers did not drive on the designated side of the road. Teachers should also encourage students to write down intermediate steps when simplifying expressions. This would improve their accuracy and therefore avoid careless errors.

Research Note: Wu (2007) was less defensive of the order of operation. According to Wu, it is unprofitable to pursue these rules vigorously in the arithmetic context the way it is done in most classrooms. He pointed out that the *left-to-right* rule is extraneous because the associative laws of addition and multiplication ensure that it makes no difference whatsoever in what order the additions or multiplications are carried out. Making the case, Wu pointed out that for young kids in the primary grades, arithmetic is essentially limited to a single arithmetic operation of between two numbers. When they confront multiple additions and multiplications such as 8 + 4 + 9 or 3 x 7 x 5, it would be wise to let them add and multiply in any order. According to Wu, they would experience the associative and commutative laws in the process. To the student, the need of accurate symbolic notation, so fundamental in mathematics when it has to be done in any level of depth, is far into the future. According to Wu, it is only around fourth grade that the idea of each arithmetic operation being *binary* (i.e., each can operate on only two numbers at a time) is brought to their consciousness and parenthesis are introduced to give explicit instructions as to which two numbers should be operated on at any point of a long succession of arithmetic operations.

Wu says that the introduction of these rules around fifth or sixth grades is either for convenience or a decision based on pedagogical considerations. His suggestion of the rule is simply, *"exponents first, then multiplications and divisions, then additions and subtraction."* He maintains that there are numerous situations where a pedantic insistence on these rules leads to bad results. His argument is that in the context of the classroom, by the sixth grade, most students already know about the associative and commutative laws of addition and multiplication and there is no reason whatsoever that students should be subjected to the stricture of performing arithmetic operations only from left to right. Moreover, they also know that subtraction is the adding of a negative number and division by x is the multiplication by 1/x.

Misconception 20

Simplifying Square Roots

Grades: 5 to 8

Question: Simplify: $\sqrt{47}$

A. 4 B. 7 C. 9 D. 10 E. None

Likely Student Answer: B. 7

Explanation of Misconception: Most students would pick 7 as the right answer by association. Because 49 is a perfect square and has a square root of 7, students would likely associate 47 as having a square root of that number. This is structural error of conceptual knowledge.

What Teachers Can Do: When it is that time to introduce students to the concept of square roots and all other roots in general, teachers should avoid starting their students off with the radical sign. All efforts should be made to give activities that would ensure conceptual understanding of square roots. One strategy is to introduce students to a geometric interpretation of square roots.

Show students pictures of three squares as shown in Figure 20.1. The edges of the first and last figures are consecutive whole numbers. The areas of all three figures are also provided. The students' task is to use a calculator to find the edge of the figure in the center. From the illustration, the students see right away that the square root of 47 cannot possibly be 7. They are able to discover that when two equal numbers are multiplied, the result is a perfect square whose square root is easily determined. They see that 6 times 6 is 36 and therefore finding the square root of 36 will produce 6; same reasoning applies to 49 as 7 times 7 is 49 and therefore the square root of 49 is 7. The teacher should point out to students that to find the edge of the side

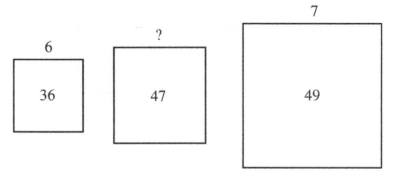

Figure 20.1. Geometric Representation of Square Roots.

of the square with area of 47, they should use methods other than using the root sign. Especially challenge students to think of two equal numbers that would produce 47 when multiplied. Obviously, some students will give 6.85, while some might even approximate to 6.9. The teacher should not zero in on the approximations and the actual answer but on the understanding of the concept. By the time students are able to relate roots and areas, they are now prepared to understand the general definition of the *nth root* of a number N as the number that when raised to the *nth power* equals N.

Research Note: The importance of multiple representations of concepts is constantly needed in mathematics. For example, the geometric interpretation of square roots in this case certainly would make the concept accessible to students. Researchers are in agreement that exposing students to multiple pathways to obtain mathematical solution is helpful for all students. For example, according to Ojose (2008), by providing mathematical representations, one acknowledges the uniqueness of students and provides multiple pathways for making ideas meaningful. Ojose states that instead of using specific algorithm to solve mathematics problems, students should be encouraged to use other methods, thus they are aware of multiple pathways of attacking the same problem. Ojose stated further that engendering opportunities for students to present mathematical solutions in multiple ways (e.g., symbols, graphs, tables, and words) is one tool for cognitive development.

Misconception 21

Comparing Negative Numbers

Grades: 4 to 8

Question: Which number is greater?
-6 or -12

Likely Student Answer: -12 is greater.

Explanation of Misconception: This student is thinking in the realm of absolute value of numbers and therefore lacks knowledge of number density. This is a structural error of conceptual understanding. The student is not demarcating between positive and negative numbers and this is typical of fifth and sixth graders. The misconception with this student has to do with thinking that the number -12 is more than the number -6.

What Teachers Can Do: Students struggle with this concept because prior to being introduced to negative integers, they did comparison of positive numbers which are absolute (e.g., 6 < 12). Comparing negative numbers could be a challenge because students erroneously attach absolute value meanings to them. It is of paramount importance to teach the concept of negative integers with models. One way is through the use of a number line. As an anticipatory set activity, a teacher could choose any integers, say, between – 50 and 50 and write one number per student on a piece of card stock. Pass out a number to each student. Have students stand and line up as quickly as possible from least to greatest without talking among themselves. After students have correctly lined up, have a class discussion about strategies that might make the process go faster next time. Shuffle the cards, pass out the numbers, and have students line up again. Even more helpful would be to have each student verbalize their number in comparison to another student's number using such expressions as "greater than" and "less than." This activity can also be modified by posting an elaborate number line on the board, assigning different numbers to students and calling on each one to come out to the board and place their number on the number line. Either way, the students will have an understanding of the relative position of each negative integer and their comparison to others. This activity should be preceded by an actual lesson in which a teacher could draw line on the board with zero in the middle. Place some positive numbers and negative numbers where they belong on the number line. Ask students to place series of numbers on the line (e.g., 0, 2, and -6) and they should talk about them. Next, they should use the < and > symbols to compare the numbers. If students are having trouble identifying the inequality symbols, point out that the symbol itself points to the lesser number. For example, in the statement -5 < 3, the inequality symbol is pointing toward -5 which is less than 3. This convention is true for all integers including zero.

Research Note: Researchers have pointed out the importance of incorporating models in teaching concepts that seem abstract for students. Providing students with models makes the concrete to abstract journey smoother for young mathematics learners. For example, Ojose (2010) discussed the various models that range from all kinds of manipulative materials to calculators. A manipulative material is a tool that assists students in advancing from concrete novice knowledge toward more expert understanding of abstract mathematical ideas and kinesthetic senses. It is worthy of being emphasized over and again that models do help children explore ideas and make sense of them (p. 119). According to Ojose (2008) hands-on materials provide students with an avenue to make abstract ideas more concrete. It allows them to get their hands on mathematical ideas and concepts as useful tools for solving problems. Many other studies have also indicated that manipulative materials enhanced learning when coupled with effective teaching techniques. For example, Raphael & Wahlstom (1989) analyzed data from the second International Mathematics Study. They found that a greater use of manipulatives was related to more comprehensive curriculum coverage, leading to higher student achievement. Furthermore, Witzel, Mercer, & Miller (2003) found that students with learning difficulties surpassed other students when taught algebra with a concrete-to-representational sequence.

Misconception 22

Addition of Negative Integers

Grades: 3 to 7

Question: What value satisfies the number sentence?
 -4 + ___ = -10

Likely Student Answer: 6

Explanation of Misconception: The student simply added 4 and 6 to obtain 10 and moved over the negative sign. This is due to lack of understanding of concepts relating to number theory, basic number sense, equations, and the equal sign. This is an arbitrary error because the student understands something but not everything.

What Teachers Can Do: If the problem above is slightly adjusted to addition of -4 and -6, chances are students may arrive at a correct answer of -10. In that case, the use of models like red counters which conventionally represents negative values may be used to solve the problem. However, the error in thinking that -4 added to 6 will give -10 is deeply rooted in the problem of transitioning from arithmetic to algebra. As a result, teachers should use models and problem solving approaches to teach the concept. Similar to the suggestions in Misconception 15, teachers should generally lead students to the idea that they can model the sum of two integers by using one color such as blue to represent positive integers and a second color, such as red to represent negative integers. With this problem, red counters can be used to represent the integers on a mat. In that case an equation is formed and the need to have equal number of counters on both sides of the equation is emphasized. It is sometimes necessary to create zero pairs for certain problems, say, 9 + (-4) understanding of which will facilitate the above problem. After creating the zero pairs, one then eliminate each pair from the expression. For example, the result of simplifying 9 + (-4) is 5 after creating and eliminating 4-zero pairs. For teachers that are introducing their students to the concept of zero pair for the first time, it is necessary to explain: *When one positive counter is paired with one negative counter, the result is called zero pair. The value of **zero pair** is 0. You can add or remove zero pairs from a mat because adding or removing zero does not change the value of the counters on the mat.* Next, the teacher should make students extend this knowledge to find an unknown. Teacher could restate the problem as 9 + __ = 5. Now that they know that -4 goes in the blank, give students similar problems to apply the knowledge of operating with integers such as this one: -4 + __ = -10. Exposing students to various forms of the same kind of problem is beneficial. Also, stating the problem in words could be helpful: *How many ripe oranges should be added to 4 ripe oranges to get 10 ripe oranges?* (Note: Ripe oranges are used here as symbolizing negative numbers. If counters or chips are used, the word problem may emphasize the red ones as per convention).

Research Note: Three research areas are identified with this concept. First, Bohan (1990) elaborated more on the use models. According to Bohan, teachers find that as concepts become more difficult, students familiar with procedures using manipulatives often get new skills as a "free ride." Second and relating to the transition from arithmetic to algebra, Filloy et al (2010) found out that extending previous algebraic syntax to the representation of the unknown requires fundamental reconceptualizations such as that of algebraic equalizations and that of mathematical unknown. Third, Baroody (1999) reported on the complementary principle of addition and subtraction in which the learning of one kind operation could in fact facilitate the learning of another operation and vice versa. Baroody presented participants of the study with items such as 4 + 5 = 9, and 9 – 4 = ?, and asked if the first item helped them to answer the second. The results showed that even though students cannot recognize they could use related addition equation to determine a difference, ready facility related addition combinations may make it more likely that children will connect their knowledge of subtraction to their existing intuitive knowledge of part-whole relations. This researcher stated that the principle leads to reasoning out differences more or less and probably slowly but with practice the reasoning process become nonconscious and automatic.

Baroody pointed out that this interpretation is consistent with the three-phase model of mental-arithmetic development long promoted by mathematics educators: informal computation, informal reasoning strategies, and mastery of basic facts. Baroody concluded that in order to facilitate mastery of the complement combinations, a teacher might first ensure mastery of related addition combinations and hope at least some children might spontaneously make the connection that this can help them answer subtraction combinations. These children can then be encouraged to explain their discovery to their peers. With those who use the complementary relation to quickly answer only subtraction combinations related to the addition doubles, a teacher might encourage a discussion to make the relation explicit and to prompt its general use. For example, a teacher could challenge children to consider whether this relation holds only for complements of the addition doubles or is generally true.

Misconception 23

Scientific Notation

Grades: 7 to 9

Question:

$$\frac{4 \times 10^8}{2 \times 10^4}$$

Likely Student Answer: 2×10^2

Explanation of Misconception: In this problem, the student divided the powers (8 by 4) to get 2 just like the student did the bases (4 by 2) to get 2 and hence the answer, 2×10^2. Understandably, students might want to know why they can divide the bases (4 by 2) but can't divide the powers (8 by 4). This misapplication of rules stems from lack of conceptual understanding of the laws guiding exponents and therefore a structural error.

What Teachers can Do: Three instructional strategies are suggested for teachers in teaching the concept of scientific notation: (a) Provide real-life contexts in which the concept of scientific notation is used, (b) Converting the numbers to decimal notations can aid the understanding and also serve as a way of checking a solution, and (c) Break up the problem and work the different pieces before reassembling. Point out to students that while it is relatively easy to represent positive numbers in conventional forms like thousands, millions, billions, and even trillions without writing the zeros, it is not so with negative numbers. For example, while it is a lot easier to write the population of California as "36 million," (we could avoid writing zeros associated with a million if we wish), we do not have that same opportunity to write in same manner extremely small numbers like the mass of carbon which is about 2×10^{-23} grams and has twenty- two zeros that comes after the decimal point. So with scientific notation, we can easily represent numbers that are very large and numbers that are very small. For these reasons that pertain to writing zeros using scientific notation, the knowledge of the concept is beneficial. Therefore, teaching it effectively so students could find the knowledge useful in their daily lives is important. Teachers could model scientific notation concepts through problem-solving approaches and strategies. For example, a teacher could make students estimate how many seconds they have been alive. With that kind of activity, students will discover that there are so many zeros and hence the need for a better representation-scientific notation. Having mastered representing numbers in scientific notation, students can then apply the concept to solving multiplication and division problems. With problems such as the one above, the teacher should encourage students to attempt to convert numbers in the problem and the answer they got to decimal forms and see if the answer is reasonable. They can confirm right away that the number 10^8 has eight zeros and the number 10^4 has four zeros and dividing them do not result in two zeros. Lastly, encourage students to break up the problems and work it in parts. They need to know that scientific notation is rewriting any number as a product of two factors and the two factors can be worked independently and then reassembled. Thus there are many routes that can be taken to learn the concept. It is therefore incumbent on the teacher to create the necessary learning tools, methods, and tasks that will maximize students' learning of scientific notation.

Research Note: Sullivan et al (2009) observed that a range of teacher actions is necessary to plan and deliver lessons utilizing potentially interesting tasks. In other words, there are specific actions that teachers must take to transform tasks into effective lessons. One of such actions is the mathematical tasks developed by the teacher. These tasks are important for teaching and the nature of students learning is determined by the type of task and the way it is used. There are some general characteristics of effective tasks. Christiansen & Walther (1986), from a mathematics education perspective, argued that non-routine tasks, because of the interplay between different aspects of learning, provide optimal conditions for cognitive development in which new knowledge is constructed relationally and items of earlier knowledge are recognized and evaluated. Ames (1992), from a motivational perspective, argued that students should see a meaningful reason for engaging in a task, that there needs to be enough but not too much challenge, and that variety is important. Fredericks, Blumfield, & Paris (2004), in a comprehensive review of studies of students engagement, argued that engagement is enhanced by tasks that are authentic, that provide opportunities for students' sense of ownership and personal meaning, that foster collaboration, that draw on diverse talent, and that are fun.

Misconception 24

Proportional Reasoning

Grades: 5 to 8

Question: A rectangle measuring 4 feet by 3 feet needs enlargement. The side measuring 4 feet becomes 7 feet in the new enlarged rectangle. What is the length of the second side? Find the area of the new enlarged rectangle?

Likely Student Answer: Second Length = 6 feet
Area of New Rectangle = Length x Width
= 7 feet x 6 feet
= 42 feet^2

Explanation of Misconception: In order to enlarge and reduce geometric shapes, students need the conceptual understanding of proportional reasoning. Obviously, this student lacks the knowledge and understanding that proportional reasoning is based on multiplicative rather than additive comparison. In this problem, the student "enlarged" the figure by adding 3 units to both dimensions (4 becomes 7 by adding 3; and 3 becomes 6 by adding 3). The student needed to realize that 4 feet became 7 feet by multiplying with 1.75 for the enlarged shape. Therefore, 3 feet should become 5.25 feet by multiplying with 1.75 (the constant ratio). Thus the enlarged shape should have dimensions 7 feet by 5.25 feet, and an area of 36.75 square feet.

What Teachers Can Do: As stated above, most students and even adults works proportional reasoning problems using additive rather than multiplicative comparison of ratios. The suggestion is for teachers to create informal activities to develop their students' proportional reasoning acumen: Identification of multiplicative relationships; selection of equivalent ratios; comparison of ratios; and scaling with ratios. For this problem, it is helpful to make students actually construct a geometric shape–like that of the enlarged rectangle. The teacher may divide the class into groups and have each group produce the enlarged rectangle that would be pasted on a pre-made poster. The enlarged rectangle should fit the pre-made one. Students should be made aware that if their rectangle does not fit the pre-made one, it means they have taken an incorrect step. Students that use additive comparison would discover that their enlarged rectangle cannot fit the pre-made one. At this juncture, teachers should point out students' misconceptions and lead them to think right. The activity can also be extended. One way to extend it is to partition the rectangle into equal parts, say two triangles each with sides 3, 4, and 5. Students could then enlarge the two triangles to form an enlarged rectangle. The rectangle can also be partitioned into other geometric shapes: a big triangle, a small triangle, and a trapezoid. Again, students should make enlarged pieces of the various shapes to form an enlarged rectangle. Always remind students that if the pieces do not fit, there is need to adjust strategy. The teacher should prompt students to think multiplicatively rather than additively. Having done this activity, teachers can now talk of the concept of constant ratio and explain its importance in working proportional reasoning problems.

Research Note: Many researchers have documented the problems that students have with proportional reasoning. For example, Pearn & Stephen (2004) noted that many misconceptions that students hold are as a result of inappropriate use of whole number thinking, including not understanding the relationship between the numerator and the denominator. They found that a major problem for students is in part because students do not understand the part/whole relationships described in fraction notation, and recommended the use of multiple representation of fractions using discrete and continuous quantities and the number line. Karplus et al (1983) noted that many of the error patterns that students demonstrate in relation to proportional reasoning problems illustrate that they apply additive or subtractive thinking process rather than multiplicative process.

Lo & Watanabe (1997) concluded in their study that certain mathematical knowledge was useful for conceptual development of ratio and proportion schemes: (a) The structure of numbers, such as divisors and multiples; (b) A conceptual basis of multiplication and division through multidigit numbers; (c) Familiarity with a variety of multiplication and division situations; (d) Meaningful multiplication and division algorithms; and (e) Integration of the above with the development of rational-number concepts. Lamon (1999) listed some characteristics of proportional thinkers: (a) They have a sense of covariation. That is, they understand relationships in which two quantities vary together and are able to see how the variation in one coincides with the variation in another; (b) They recognize proportional relationships as distinct from non-proportional relationships in real-world contexts; (c) They develop a wide variety of strategies for solving proportions or comparing ratios, most of which are based on informal strategies rather than prescribed algorithms; and (d) They understand ratios as distinct entities representing a relation different from the quantities they compare. Along the same line of argument, Van de Walle (2007) says that proportional thinking is developed through activities involving comparing and determining the equivalence of ratios and solving proportions in a wide variety of situations and contexts (p. 353).

Singh (2000) evaluated ninth grade students' understanding of proportional reasoning on missing value, numerical comparison, and qualitative reasoning tasks to determine if a relationship existed between national exam scores and proportional reasoning tests. Results showed that in missing value problems, most students used additive rather than multiplicative reasoning. In numerical comparison tasks, "how much for one?" was the most common interpretation, and in qualitative reasoning tasks, students typically failed to construct inverse relations or to take all the relevant factors into consideration when attempting to reason qualitatively. Singh found that students had an instrumental understanding, rather than a relational understanding, of missing value tasks. As a result, students were unable to apply their knowledge to new contextual situations. Singh recommended that (1) sense-making be central in teaching proportional reasoning, (2) numerical comparisons, qualitative, and nonroutine problems be included in lesson plans, and (3) proportional reasoning assessments be consistent with teaching and learning emphasizing meaning.

Misconception 25

Time Problem

Grades: 2 to 6

Question: Your family trip from Las Vegas started at 1:55 pm. Your arrival time in Los Angeles was 7:08 pm. How long was the trip?

Likely Student Answer: It took my family 5 hrs and 53 minutes as follows:

$$\begin{array}{r} 7:0\,8 \\ -1:5\,5 \\ \hline 5:5\,3 \end{array}$$

Explanation of Misconception: This misconception is associated with misapplication of the subtraction algorithm to time problem. Rightly, the student felt it necessary to subtract 1.55 from 7.08. However, the student did not take into consideration that the problem involve time in which 60 minutes represent 1 hour, and therefore ought to have regrouped with 60 minutes for 1 hour instead of 100 minutes for 1 hour when performing the subtraction operation. Done correctly, student should have arrived at 5 hours 13 minutes.

What Teachers Can Do: Problem solving exercises such these are important part of the K-8 curriculum. However, elementary and middle school students often have difficulties with these ideas. Adding and subtracting time involve understanding the relationships between minutes and hours. Two difficult situations may be encountered by students in this problem. One is the fact that the question is problem-centered and in which case the information needed to solve the problem has to be extracted. Two is the problem with actually carrying out the subtraction algorithm. See Misconception 2 for strategies related to subtraction. One key skill needed for problem-solving is to read and understand what the question demands before beginning the work. Also, the units to use with the regrouping which obviously was a problem for the student need to be addressed. To mitigate this problem, it is suggested that teachers use a clock as a physical model. The clock will show students practically the total number of hours and minutes the journey lasted. The clock is also a good model to show students that 1 hour represents 60minutes, not 100 minutes as exhibited by the student. Another strategy is to use the distance-between-times model: from 1:55pm to 2:55pm is 1 hour, therefore from 1:55pm to 6:55pm is 5 hours; and from 6:55pm to 7:08pm is 13 minutes. Therefore the journey took 5 hours and 13 minutes. If a teacher does this first, by the time the teacher eventually begin to explain the concept using subtraction algorithm, it will make more sense to the students.

Research Note: Research by Thanheiser (2009) focused on time task problem concluded that both students and even preservice elementary school teachers (PST) experience difficulties with the concept. This researcher presented PSTs with a student's incorrect application of the standard subtraction algorithm (for base-ten numbers) to find elapsed time, an inappropriate context for this algorithm, and were asked to comment. PSTs who recognized that the child was incorrect (in regrouping 100 instead of 60 minutes for 1 hour) were asked to explain the error and to alter the algorithm to make it applicable in the time context (regroup 1 hour into 60, rather than 100 minutes). Those who thought that the child was correct were asked to solve the problem in an alternative way – of course the child's solution was wrong because he incorrectly applied the standard subtraction algorithm for base-ten numbers.

Results of the study showed that six (40%) of the 15 PSTs in the study immediately recognize that the standard subtraction algorithm did not apply in this situation; 9 PSTs (60%) initially thought that the child was correct, but 8 of these 9 changed their minds after they had calculated the correct time difference another way. Even though 14 of the 15 PSTs eventually stated that the child's solution was wrong, only 5 explained both the mistake the child made (in regrouping 100 instead of 60) and a way to alter the algorithm to make it applicable in a time satiation. Four of the 5 PSTs whose conceptions were categorized as either reference units or groups of ones in the context of the standard algorithms immediately recognized that the application of the standard algorithms in a time situation is incorrect, and all 5 were able to alter the algorithm to make it applicable. Of the 10 PSTs with the two other conceptions in the context of the standard algorithms, 2 (20%) immediately recognize that the standard algorithm did not apply in a time situation and 3 (30%) were able to explain the mistake and alter the algorithm as needed. The researcher concluded that to help their students learn about numbers and algorithms, teachers need more than ability to perform algorithm. They need to be able to explain the mathematics underlying the algorithms in a way that will help students understand.

Part Two

ALGEBRA

Part two of this book is focused on misconceptions that students encounter in algebra. Students in middle and high schools are often engrossed in misconceptions when transitioning from arithmetic to algebra. This transition which usually involves manipulation of symbols and expressions differs from simple arithmetic processes. These misconceptions poses major challenge to students and attempt by teachers to address them could help in the learning of more complex concepts of mathematics in the future. Dominant in the study of algebra are concepts like equations, inequalities, algebraic expressions, factoring expressions, systems of equations, polynomials, exponents, and exponential equations. The part two of the book is devoted to these concepts.

Misconception 26

Dividing Rational Expressions

Grades: 7 to 12

Question: Simplify

$$\frac{X+5}{5}$$

Likely Student Answer:

$$\frac{X+5}{5} = X$$

Explanation of Misconception: This misconception is due to students' inability to recognize that only factors can cancel out. Stated differently, students need to know that in order for any number, variable, or expression to cancel out, there must be common factors in the numerator and denominator. In this case above, the numerator and the denominator have no common factors; so the expression cannot be simplified. One reason that could possibly explain the cancellation error by students is that they think that every expression is subject to further simplification and an answer ought to be produced, regardless.

What Teachers Can Do: Two approaches are suggested for teachers to help students with this misconception: (a) construct tables of values to point out error that students are making and (b) show the two expressions on a graph to see if they are same or different. With (a), the teacher could help students

Values to Substitute	$\dfrac{X+5}{5}$	X
3	(3+5)/5 = 1.6	3
5	(5+5)/5 = 2	5
10	(10+5)/5 = 3	10
15	(15+5)/5 = 4	15

Figure 26.1a. Table of Values for $\dfrac{X+5}{5}$.

Values to Substitute	$\dfrac{5 \cdot x}{5}$	X
3	(5•3)/5 = 3	3
5	(5•5)/5 = 5	5
10	(5•10)/5 = 10	10
15	(5•15)/5 = 15	15

Figure 26.1b. Table of Values for $\dfrac{5 \cdot x}{5}$.

see a reason why their approach is wrong and to correct the error by having them make two tables of values. One of the tables should be constructed using the unsimplifiable expression that is given above: $\dfrac{X+5}{5}$. The other table of values should be constructed using an expression in which the terms can simplify (cancel out).

As shown in Figure 26.1a, for every given value of X, the corresponding value of the expression $\dfrac{X+5}{5}$ is different. Make students think about possible reasons why this is so. Next, give students an expression that could be simplified like: $\dfrac{5 \cdot x}{5}$ and have them make a table of values as shown in Figure 26.1b. How is this table different from the previous one? Students should discuss the two tables.

Students will see that as a contrast to table 1a, this second table shows that for every given value of X, the corresponding value of $\dfrac{5 \cdot x}{5}$ is same. Hence it okay to cancel out the 5's because they are common factors. But this is not the case with the expression $\dfrac{X + 5}{5}$ which has no common factors. With (b), if the teacher has taught the unit on graphing linear equations, this will be an appropriate time to reinforce that concept by showing students that for their reasoning to cancel out to be correct, the graphs of y = X, and y = $\dfrac{X + 5}{5}$ must be the same. Have students produce the two the graphs. They will discover that the graphs are different and therefore the expression $\dfrac{X + 5}{5}$ does not simplify to X.

Research Note: Researchers have identified student's overreliance on visual cues as a reason for these kinds of misconceptions. The reliance on visual cues therefore leads to cancellation error. A visual cue according to Demby (1997) is anything in the structure of a mathematical object that brings to mind a piece of mathematical knowledge or a memory of mathematical experience. Demby stated that reliance on visual cues is not necessarily undesirable since masterful algebra students can use visual cues efficiently, appropriately, and successfully. However, visual cues may lead many students to commit errors by inappropriately canceling for example, 3 and -3 appearing in the expression 2x – 3 + 3x because they are visually attractive. Kirshner (1989) found that some students rely on visual cues rather than propositional reasoning, and students who lack a propositional foundation for their syntactic knowledge often retrieve inappropriate rules. Both researchers noted that the inappropriate use of visual cue lead to *cancellation error*. They gave an example of students canceling 3x from the numerator and denominator of (3x + 6) / 3x because they were prompted by recalling the cancellation property of (ab)/a = b. The researchers pointed out that it is important to note that a student committing such a cancellation error may indeed possess the corresponding propositional knowledge but may simply have failed to activate this knowledge in the face of such strong visual cues.

Misconception 27

Adding Rational Expressions

Grades: 7 to 12

Question: Add

$$\frac{1}{x} + \frac{1}{y}$$

Likely Student Answer:

$$\frac{1}{x} + \frac{1}{y} = \frac{2}{xy}$$

Explanation of Misconception: This misconception relates to generalizing procedures previously learned with whole numbers to rational expressions. The student thinks that adding the numerators which are numbers and multiplying the denominators which are variable suffices for the problem. This is a structural error associated with lack of conceptual understanding.

What Teachers Can Do: This error can be corrected by giving a similar problem in arithmetic and having students relate it to algebra. For example, what is 1/5 + 1/6? Will the solution to this problem be 2/30 or 2/11 or some other fraction? Challenge students to think again about their pie or circle graphs. After coming up with the right solution to 1/5 + 1/6, ask students to compare it with the rational expression problem: 1/x + 1/y. The teacher should point out and emphasize that the same rule apply to fractions and rational expressions. Having given students the opportunity to think *right*, the teacher could give a similar problem on the board and employ the following lesson study approach: (a) make students write about the process of getting the answer and explain it to other members of the class; (b) make students critique the work of other students; and (c) have students correct the mistake of other students. By having students write about their solution sequence, they are reflecting on what they did and how they did it thereby reinforcing the concept in meaningful ways. The teacher could select a few students to read out their write-up to the rest of the class. Also, by having students critique each other's work, they have an opportunity to engage in an exercise that puts them in expert position. Students are empowered if they see themselves as experts. Remind students that critiquing work is not the same as criticizing. If possible, model the process and let them become comfortable before they practice. With this problem, it is important for the teacher to present to students some kind of systematic way of recognizing the transition from arithmetic to algebra. For example, having seen the relationship between $1/5 + 1/6$ and $1/x + 1/y$, the schema needed by students to develop the connection should constantly be reinforced. A major obstacle with rational expressions is that it may be the first entity in algebra for which students are not given or are able to generate a concretization to correspond with the expression. Teachers should acknowledge this reality and plan accordingly.

Research Note: With this topic, two areas of research are identified: Problems associated with transitioning from arithmetic to algebra and strategies including those consistent with lesson study approach. Demby (1997) did much work on students' transition from arithmetic to algebra. Demby noted that skills and meaning should somehow develop together and should reinforce each other. Each student should construct their own meaning of algebraic transformations, even though different from ours. Demby also pointed out that construction of the child's knowledge does not follow from deductive reasoning, but from gathering experience, comparing results, and generalizing and revising their previous rules. Therefore constant verification of what the students do is crucial. According to Demby, learning algebraic transformations by practicing without attempt to generalize the child's experience is futile and that children cannot remember specific tasks but can remember certain conclusions, especially those that surprised them by contrast to their prior beliefs. Demby also states that transmitting algebraic rules by the teacher, memorizing them by students and practicing them in a mechanical way is not helpful.

An account of arithmetical roots of the early algebra was given by Kieran (1992). This researcher pointed out that the traditional emphasis in the curriculum on "finding the answers" allows arithmetic learners to get by with informal, intuitive procedures; however, in algebra they are required to recognize the structure they have been able to avoid in arithmetic. One of the pressing factors for algebraic reform is the ability of elementary teachers to "*algebrafy*" arithmetic (Kaput & Blanton, 2001), that is, to develop in their students the arithmetical underpinnings of algebra (Warren & Cooper, 2001). As was argued by Carpenter and Levi (2000), the artificial separation of arithmetic and algebra "deprives children of powerful schemes of thinking in the early grades and makes it more difficult to learn algebra in the later years" (p. 1). However, the popular belief that algebra should be introduced as growing out of arithmetic was challenged by Lee and Wheeler (1989). They gathered test-interview data which showed that such an approach faced serious pedagogical difficulties and might obscure the genuine obstacles. It was found that for many students (even some of those who were successful at standard algebra tasks) algebra and arithmetic were two dissociated worlds; when these students were confronted with both, their arithmetic appeared disturbed by algebra.

With lesson study, researchers are in agreement with regard to its efficacy. For example, Kelly (2002) reviewed the implementation of lesson study as a method for systematically improving clarion instruction by replacing teachers' ingrained assumptions and solo practices with collaborative brainstorming, planning, observation, and evaluation. Kelly noted that implementation of lesson study typically consisted of systemic steps aimed at student success and should be focused on long-term improvement goals.

Misconception 28

Adding Unlike Terms

Grades: 6 to 9

Question: Add: $2X + 4 + 3X + 5$

Likely Student Answer: $14X$

Explanation of Misconception: This misconception is notation-related. The student combined like and unlike terms incorrectly because of the erroneous thinking that operation symbols cannot be part of an answer. The student did not recognize that operational symbols, in this case the plus sign, ought to be in the answer: $(5X + 9)$.

What Teachers Can Do: Three strategies could be helpful in teaching this concept: Use of counter examples; use of real-world problems; and the use of models. Because students do not recognize that an operation symbol could be in an answer, it is highly recommended that the use of models is central to teaching the concept. A teacher may begin a lesson by asking questions that would act as anticipatory set for the lesson: "*If you add three apples to four oranges, how many things do you have in all?*" Chances are no student will say "*seven oranges*" or "*seven apples.*" Responses might include, "I don't know" or "There is no answer for it" or "It cannot be added" or even "7 apples and oranges." The teacher could use students' responses as a spring- board to start teaching the concept. The teacher should state in the affirmative that just like it is not possible to have "*7 apples*" or "*7 oranges*" as answers to the above stated word problem, there are expressions that cannot be simplified or can only be simplified to a certain extent. In the context of this problem, the teacher could say something to the effect that the numbers in the problem that has X's could be likened to number of apples and the stand-alone numbers (numbers without variable X attached to them) could be likened to the oranges. Therefore, it is okay to take the apples and put them together; likewise, it is okay to put the oranges together (combining). After combining, they should perform the operations and write out an expression. This will show that there are five apples and nine oranges. Remind students that they have previously been apprised that apples and oranges could not be "combined" together and so we are left with a statement or expression of "five apples and nine oranges" which is algebraically written as 5X + 9. The teacher should emphasize the importance of leaving the answer in this form with the operation symbol in it.

Another good example is money. Ask students to come up with a total amount of currency using two different countries. For example: *Ten U.S. dollars plus 5 Mexican pesos equal what amount of money?* The more the teacher relates this kind of concept to real-life contexts, the more students will grasp the concept. Models like algebra tiles should also help a great deal. The teacher could use the X-tiles and unit tiles to do *combining* like terms. The teacher should also incorporate drawing of the models to illustrate the concept. For example:

If 2X is represented with Q Q,

And 4 represented with Ò Ò Ò Ò,

Then 3X = Q Q Q

And 5 = Ò Ò Ò Ò Ò

Therefore, 2X + 4 + 3X + 5 = Q Q + Ò Ò Ò Ò + Q Q Q + Ò Ò Ò Ò Ò

Rearranging, we have: Q Q + Q Q Q + Ò Ò Ò Ò + Ò Ò Ò Ò Ò

This is same as: 2X + 3X + 4 + 5

Which is: 5X + 9

Other than algebra tiles, the teacher could use any other kinds of models ranging from chips to cubes in teaching the concept. But be sure to assign specific meaning to the objects being used. If algebra tiles are used, then the conventional assignment of X to the rectangular shaped tile and assignment of a unit to the small square tile are appropriate.

Research Note: Researchers have distinguished between semantic and syntactic aspects of algebra. Demby (1997) noted that semantic aspect is understood as generalized arithmetic while syntactic refers to the way an expression can be manipulated according to formal rules regardless of symbolic meanings. Demby further categorized some attributes of algebraic processes and problem solving into a semantic-syntactic dichotomy: Automatization and Quasi Rules are both semantic and syntactic; Preparatory Modification and Guessing-Substituting are semantic; Concretization is semantic; and Rules and Formulas are syntactic. In their transition from arithmetic to algebra study, Lee & Wheeler (1989) remarked that the track leading from arithmetic to algebra is littered with procedural, linguistic, conceptual and epistemological obstacles. They quoted Filloy and Rojano as having suggested that the challenge of dealing with syntax of an unfamiliar algebraic language tends to destabilize students' semantic control of the familiar arithmetical language. Tall & Thomas (1991) and Gray & Tall (1993) pointed out that with respect to the syntax aspect of algebra, those children who can interpret arithmetic symbols only as operations to be carried out are confused by algebraic expressions such as 3a + 4b.

Related to the use of concrete models in teaching this concept, Filloy & Sutherland (1996) identified two fundamental components: *translation and abstraction*. Translation encompasses moving from the state of things at a concrete level to the state of things at a more abstract level. The model acts as an analogue for the more abstract. Abstraction is believed to begin with exploration and use of processes or operation performed on lower level mathematical constructs (English & Sharry, 1996). Through expressing the nature of their experience and articulating their defining properties, learners can construct more cohesive sets of operations that subsequently become expressions of generality. It is conjectured that abstraction entails recognizing the important relational correspondence between a balance model, weights as numbers, stating the identity of the two quantities or expressions, and applying the understanding to more abstract situations, such as, algebraic expressions. Filloy & Sutherland suggests that not only do models often hide what is meant is to be taught, but also presents problems when abstraction from the model is assigned to the students. Teacher intervention is believed to be a necessity if the development of detachment from the model to construction of new abstract notion is to ensue.

Misconception 29

Adding Like Terms

Grades: 4 to 8

Question: 3X + 3X

Likely Student Answer: $3X + 3X = 6X^2$

Explanation of Misconception: This misconception may be connected to the erroneous thinking that while numbers are added, variables should be multiplied to simplify an expression. The student applied inappropriate rules and therefore a structural error related to conceptual understanding. In comparison to problems involving unlike terms, this problem has like terms and simplifying it does not involve elaborate algebraic manipulations.

What Teachers Can Do: Students may sometimes apply inappropriate rules, especially if the rules are confusing or if the rules do not make sense to them. The use of models is crucial in helping to address these problems. Teachers are encouraged to use models like chips, algebra tiles, counters, cubes or others available to them in teaching this concept. It is advisable that students physically move the objects being used. For example, objects like apples could be assigned thus:

3X = Ŏ Ŏ Ŏ

3X = Ŏ Ŏ Ŏ

3X + 3X = Ŏ Ŏ Ŏ + Ŏ Ŏ Ŏ

= 6X

In addition to models, the teacher should encourage students to list out terms. For example,

$3X = X + X + X$

$3X = X + X + X$

Therefore, $3X + 3X = X + X + X + X + X + X = 6X$

Teachers should not look at these representations as too easy for students and therefore unnecessary. They should form the repertoire of a teacher's instructional strategy as students conceptually grasp the ideas through these illustrative processes. These kinds of representations are necessary before introducing students to rules and algorithms. Also, teaching the concept contextually is important. For example, real-life situation could be used to model this problem: *There is 3 dollars in your mother's closet and there is 3 dollars in your father's closet. How much is that in all?* The use of words could make students become comfortable and would assist them in dealing with symbolic expressions in general.

Research Note: A major problem associated with this concept involves the transition from arithmetic to algebra. Some researchers and professional organizations have supported the early exposure of students to algebra as a way of mitigating the problem. For example, the stance of the National Council of Teachers of Mathematics (NCTM, 2000) is that algebra in K-12 be a continuous strand. According to NCTM, "By viewing algebra as a strand in the curriculum from pre-kindergarten on, teachers can help students build a solid foundation of understanding and experience as a preparation for more sophisticated work in algebra in the middle grades and high school" (p. 37). The impact of this reconceptualization has been most apparent in elementary school, where mathematics educators have recently made concerted efforts to integrate algebraic ideas (Kaput, Carraher, & Blanton, 2007). Integration of algebra in the elementary grades implies recognition of algebraic reasoning as more than of equation manipulations and viewing algebra as including new forms of reasoning accessible to students across the grades (Carpenter & Levi, 2000).

A great deal of evidence has shown that young children develop an intuitive and informal sense of mathematical concepts and procedures before they enter school (Hughes, 1981; Starkey & Gelman, 1982). Educators agree that there is need to capitalize on these informal understandings and find ways to help children relate them to the formal mathematics taught in school (Wood & Sellers, 1997). However, not all researchers agree with early exposure. For example, Asquith et al (2007) noted that introducing algebraic ideas to students earlier present many challenges, including learning more about the development of students' early algebraic reasoning, designing supportive curricula, and developing teacher knowledge and practice that will enable teachers to foster connections between arithmetic and algebraic forms of reasoning. These challenges are particularly relevant at the middle school level, at which time the transition from arithmetic to algebra thinking is arguably most salient, they opined.

Misconception 30

Distributive Property

Grades: 4 to 9

Question: Simplify
$$3(x+y)$$

Likely Student Answer: $3(x+y) = 3xy$

Explanation of Misconception: This misconception is associated with misapplication of rules. The student must have learned that they needed to multiply the outside term by each of the terms in the parenthesis, and hence the multiplication of all three terms without regard to the operation symbol. Somehow the student did not recognize a correct procedure in which the outside term is multiplied by each term in the parenthesis while also maintaining the operational symbol (+).

What Teachers Can Do: Teachers should concentrate on the big idea of the distributive property. The distributive property says that if we take the sum of some numbers and multiply it by some other number, the result is the same as if we first individually multiplied each number in the sum and then added them all up. For example with $6(2 + 3)$, you can either add first and then multiply by 6 or multiply $6(2)$ and $6(3)$ first before finding the sum. It is important to highlight that the distributive property could be used in different situations: numerical expressions only; a mixture of numerical and variable expressions; and variable expressions only. For example, we could have $3(2 + 6)$, $3(x + y)$, and $z(x + y)$. These are all expressions in which the distributive property can be performed. While the first one is arithmetic, the second and the third are algebraic. A suggested strategy is to expose students to numerical problems involving the distributive property and thereafter transition to the algebraic ones. By way of helping students understand the concept of the distributive property more deeply, a teacher could give the following factor partitioning activity: Supply students with several sheets of centimeter grid paper. Assign each pair of students a product such as 6 x 8. (Product can vary across the class or all be the same.) The task is to find all of the different ways to make a single slice through the rectangle. For each slice the students should write an equation. For example, for a slice of one row of 8, students would write: 6 x 8 = 5 x 8 + 1 x 8. This can be repeated for several other numbers as time permits. Next, the teacher could use geometric figures to illustrate the concept of the distributive property.

Have students draw the shapes as shown in Figure 30.1 and label them appropriately. The first illustration is a two-in-one rectangular shape in which the area needs to be determined by simply multiplying the width and the length. Tell students that for the purpose of clarity, they should assume that there will be a "left side" and a "right side" of the distributive property. Remind students that the total length of the shape is (2 + 6) which is 8; and a width of 3. This gives an area of 24 and should be regarded as the "left side" of the distributive property. For the right hand side, the two shapes could be separated and the areas found thus: $3(2 + 6) = 3(2) + 3(6) = 6 + 18 = 24$. The students would therefore see that one side of the distributive property is equal to the other side. More importantly, they see the connection that $3(2 + 6) = 3(2) + 3(6)$. At this point, the teacher should make reference to the second illustration that has variables. The teacher could have students brainstorm on the similarities and differences between the two figures. The teacher could conclude by reminding students that just like we have in arithmetic, we could also represent the distributive property in algebra: $3(x + y) = 3(x) + 3(y) = 3x + 3y$.

Research Note: Students' difficulties in learning the basic ideas of early algebra have been well documented. Kieran (1989) emphasized that an important aspect of this difficulty is students' lack of ability to recognize and use structure. Structure according to Kieran includes the "surface" structure (e.g., that the expression $3(x + 2)$ means that the value of x is added to 2 the result is then multiplied by 3) and the "systemic" structure (the equivalence forms of

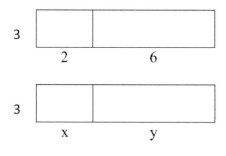

Figure 30.1. Geometric Representation of the Distributive Property.

an expression according to the properties of operations). Kieran sees algebra as the formulation and manipulation of general statements about numbers, and hence hypothesizes that children's prior experience with the structure of numerical expressions in elementary school should have an important effect on their ability to make sense of algebra. Booth (1989) expresses the same view, ". . . a major part of students difficulties in algebra stems precisely from their lack of understanding of arithmetical relations. The ability to work meaningfully in algebra, and thereby handle the notational conventions with ease, requires that students first develop a semantic understanding of arithmetic . . ." (p. 58).

Vermeulen et al (1996) did some extensive work on the distributive property. They designed learning activities for teaching of the distributive property concept and tested the efficacy of their activities which included the use of concrete problems, decontextualised problems, and the problems that encourage decomposing the numbers to apply the distributive property. Students in grades 3, 4, 6, and 7 were pretested and posttested. Results indicated that grades 6 and 7 experienced a substantial increase in their awareness of the distributive property. The outcomes for grades 3 and 4 were however, very inconsistent, leading them to conclude that their strategy did not benefit these students in acquiring a higher level of awareness of the distributive property. According to the researchers, a possible explanation for this is that the younger students found the calculations in the tests very complex–their thought processes were occupied in trying to manage the calculations and not towards reflections of these thought processes. According to the researchers, using calculators in the tests may change the outcome of the results.

Misconception 31

Writing a Variable Expression

Grades: 5 to 8

Question: Write the variable expression for "X less than 6."

Likely Student Answer: X - 6

Explanation of Misconception: This s a common error usually made by English Language Learners (ELL) because the question requires decoding information from a word problem. A skill required for this kind of problem includes the ability to correctly read and interpret information. This student does not obviously possess that skill. The student exhibited confusion with the positions of the X and the 6.

What Teachers Can Do: Because the knowledge of word problem (used interchangeably with problem solving and story problem) is key in mathematics, teachers should recognize their students' difficulties in the area and plan instruction accordingly. It is important to note that understanding what a question requires is a critical component in devising a solution strategy. Also worthy of note is that majority of the questions that students encounter in standardized tests are stated in words – the more reason why teachers should pay attention to them in instruction. In general, the following tips are important when teaching problem solving. First, word problems should not be taught in isolation. For example, the above problem should be taught when a teacher is on the unit of variable expressions. This will help avoid a situation where students see concepts stated algebraically as different from concepts stated in words.

Second, incorporating children literature in teaching mathematics is important. This makes the concept relevant to the students. Children's stories can be used in numerous ways to create reflective tasks at all levels. By way of example, a very popular children's book, *The Doorbell Rang* (Hutchins, 1986), can be used with different agendas at various grade levels. The story is a sequential tale of children sharing cookies. On each page, more children come to the kitchen and the same number of cookies must be redistributed. This simple but yet engaging story can lead to exploring ways to make equal parts of almost any number for children at the K-2 level. It is a spring-board for multiplication and division at the 3 – 4 level. It can also be used to explore fraction concepts at the 4 – 6 level. One benefit of this approach is that students' skills associated with decoding information in a problem are enhanced. Also, the process affords students multiple entry points to a problem, particularly ELL students. It is suggested that teachers find and use books that can make content clearer to students.

Third, teachers should insist on social conventions of symbolism and terminology that are important in mathematics and that cannot be developed through reflective thought. For example, representing "seven and five equals twelve" as "7 + 5 = 12" is a convention. Definitions and labels are also conventions. What is important is to offer these symbols and words only when students needs them or will find them useful. Lastly, general strategies of problem solving are recommended: draw a picture, act it out, use a model, look for pattern, make a table or chart, make a list, and possibly try a simpler form of the problem.

Research Note: A number of studies have shown that it is important to engage students in problem solving as that practice promotes conceptual understanding. For example, Woods & Sellers (1997) assessed a problem-centered mathematics program. They did a longitudinal analysis of mathematics achievement of three groups of elementary students: those who received years of problem-centered instruction, those who received one year of problem-centered instruction, and those who received textbook instruction. Their findings showed that after 2 years of problem-centered classes, students have significantly higher achievement on standardized achievement measures, better conceptual understanding, and more task-oriented beliefs for learning mathematics than do those in textbook instruction. In addition, these differences remain after problem-centered students return to classes using textbook instruction. Also, on measures developed to assess students' conceptual understanding of arithmetic, the results of the analyses indicate that students in problem-centered classes for 2 years score significantly higher on the Relational Scale than textbook-instructed students in both third and fourth grades. In addition, students in problem-centered classes scored significantly higher on majority of the subscales, specifically on those that were developed to measure students' understanding of numeration, place value, and multiplicative situations.

Ward (2005) reviewed the research and resources that support the integration of literature into the teaching and learning of mathematics. Ward found that literature had a unique advantage in the mathematics classroom because many abstract mathematical concepts could be presented within the context of a story, using pictures and more informal language. Ward also found that by integrating mathematics and literature, students gained experience with solving word problems couched in familiar stories and thus avoided struggling with unfamiliar vocabulary. To make the learning of mathematics more engaging and less intimidating, Ward's review emphasized that integrating literature into mathematics curriculum provided a natural setting for observing mathematics in real world, making it come alive, and thus conveying real meaning to students. Ward concluded that teacher educators should aim for producing teacher candidates who are literature savvy and prepared to teach mathematics with a focus on reading, writing, and communicating mathematically.

Misconception 32

Simplifying a Variable Expression

Grades: 8 to 12

Question: Consider the following expression: $X + Y + XY$

(a) How many terms are in the expression?

(b) Simplify the expression

Likely Student Answer: (a) There are four terms

(b) $X + Y + XY = 2X2Y$

Explanation of Misconceptions: (a) The student counted two X's and two Y's to arrive at the four terms for the expression. The student did not consider the last term (XY) as one term. (b) The student extended the notion of two X's and two Y's to arrive at 2X2Y. These are structural errors that are related to conceptual understanding. The conceptual knowledge of variables and coefficient is lacking. The student thinks that "simplify the expression" means that they must arrive at an answer regardless of whether the expression could be simplified or not.

What Teachers Can Do: In general, students have difficulties with the concept of polynomials. The mathematical language associated with the concept is to say the least alien to students. This is especially true if the course is students' first algebra course. Teachers could do three things: (a) create activities geared towards helping students recognize the number of terms in polynomials; (b) teach vocabulary associated with unfamiliar terms (e.g., monomial, binomial, and trinomial, etc); and (c) use models to teach the concept. First, a teacher could create index cards that have the expressions written on the front and the corresponding number of terms on the back side.

Misconception 32

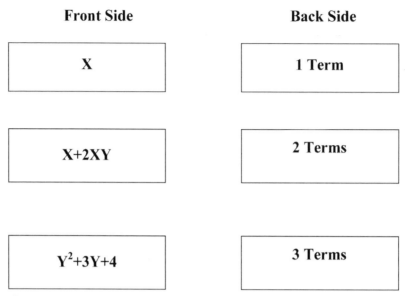

Figure 32.1. Terms Matching of Variable Expressions.

These could be used as reinforcement activities or even as warm –up activities before instruction begins. For example, on a card that has "X" on the front side, "1" should be on the back side; a card that has "X + 2XY" should have "2" on the back side; a card that has "Y^2 + 3Y + 4" should have "3" on the back side, and so on. The teacher should hold up a card at a time and call on students to say the number of terms that is displayed on the card. After a response, the teacher could then reveal the number of terms by turning the card to the flip side. After students are familiar with the general idea of the cards, the teacher could involve them in future card making process.

In addition to the above strategy, the teaching of vocabulary should also be a focal point of the unit. As mentioned earlier, transitioning from and arithmetic to algebra and the attendant plethora of new vocabulary that comes with it could be overwhelming for students. Teachers should recognize this fact and make efforts to get their students familiar with the vocabulary. Explicit instruction in teaching new vocabulary terms requires the teacher to directly teach the pronunciation and definitions of new vocabulary word in a highly specialized manner. In the area of algebra, it also requires teachers to provide students with examples and problem types. For example, in trying to help students understand the word "binomial," a teacher could first of all point out the "bi" in the word and have them give feedback on what the word could mean. Chances are students may associate it to "two things." The teacher can then make students come up with other words starting with "bi." Chances of mentioning "bicycles" or "biracial" or "bisexual" are quite high (I have done this in the past and students' responses are always similar to these). This is the point where a teacher will say that "binomials are polynomials with two terms." The teacher should then give examples of binomials such as $3 + X$, $5X^2 + 2$, $6X - Y^2$, etc. Lastly, the teacher should use some kind of model to illustrate the concept to students. Algebra cubes are three-dimensional and are appropriate in teaching polynomials.

Research Note: One way to increase students' vocabulary is to pre-teach vocabulary terms before students encounter the words in their mathematics textbook or during instruction. When pre-teaching vocabulary, it is important to teach a word within its context. Carnine, Silbert, & Kameenui (1997) suggest the following methods for teaching new vocabulary words: State the definitions and have students repeat definitions; provide students with good examples of the word; and review new words along with previously learned words to ensure students have the words in their long-term memories. Once students have explicitly been taught the new vocabulary terms, they can continue practicing the words by using self-correcting cards. Other pronouncements by researchers such as Chard (2003); Mercer, Lane, Jordan, Allsopp, & Eisele (1996); and Vaidva (2004) are as follows: (1) Solving word problems is dependent on understanding the language in the word problems; (2) To avoid memory gaps and misunderstanding, vocabulary should be explicitly pre-taught and reviewed before each new math lesson; (3) Students often require examples illustrating the context of vocabulary words, and they should be encouraged to use the words in journals, presentations, and explanations of work; (4) When learning is difficult and novel, teachers must provide support for their students; (5) Teachers must model what they want students to learn and provide guided instruction, independent practice, and frequent feedback; and (6) Students must be given ample opportunities to practice the task in order to generalize the strategy for other settings.

Misconception 33

Factoring

Grades: 7 to 9

Question: Factor
$$\pi y^2 + 4y^2$$

Likely Student Answer: $\pi y^2 + 4y^2 = 4\pi y^2$

Explanation of Misconception: Again, this student felt it necessary to just "combine" all terms. This is a structural error as the student has no clue as to what the problem required. The student probably handled it this way because of the addition symbol which the student misconstrues as a command to find an answer. This expression is not factored by simply multiplying the coefficients of y^2. The problem requires factoring out a term that's common to both πy^2 and $4y^2$. The common term is y^2 and factoring out that common term would leave an answer of $y^2(\pi+4)$.

What Teachers Can Do: Teachers should teach this concept with methods that are non-traditional. Traditional methods whose routine is limited to explaining concepts and then giving students work to do afterwards does not help students learn conceptually. For example, teaching this concept through activities of geometrical representations may be more beneficial than just explaining the concept to student without any activity to support the learning. The teacher will capitalize on the prior knowledge relating to the area of rectangles that students already possess.

First, students can relate to the fact that the combined area of the two shapes would involve area of A + area of B which is $\pi y^2 + 4y^2$. Also, students should be able to acknowledge that Area of A and Area of B do have something in common. Obviously, they have common width of y^2. Therefore, calculated together, we would have the common width (y^2) multiplied by

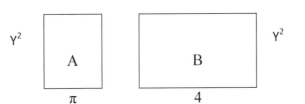
Figure 33.1. Factoring through Area Model.

the total length $(\pi + 4)$ to obtain $y^2 (\pi + 4)$. Thus we can write: $\pi y^2 + 4y^2 = y^2 (\pi + 4)$. Expressions on both sides of the equation are alternative ways of writing the same thing. It shows that factoring is the reverse of distributing multiplication over addition or subtraction. At this point, the teacher can now talk of factoring out a common term from an expression just as shown in this case. By way of checking for understanding, the teacher could give similar problems and see how well the students do. The teacher could also use physical models like algebra tiles to teach this concept.

In general, the lack of conceptual understanding exhibited by students in factoring and finding product of polynomials is due to the use of teaching methods that students do not connect with. One of such methods is the FIOL method which seems to be too mechanical as it involves rules and steps of order of multiplication and addition. In the case of factoring an expression such as $X^2 + 5X + 6$, the teacher may give several steps in working the problem with FIOL. Such steps are not only hard to remember but are changed from one form of expression to another. For example, the steps for factoring $X^2 + 5X + 6$ would be significantly different from the steps required for factoring $2X^2 + 7X + 3$. Thus the use of models like algebra tiles is highly recommended. Algebra tiles can be used to physically illustrate how the x^2- tile, the x-tile, and the unit tiles form a rectangle whereby the length of sides represent the factors of the expression. In this dispensation, the prior knowledge of students is important. The teaching should be focused on conceptual understanding through constructive methods rather than on *routinized* methods that emphasize steps, explanations, and procedures.

Research Note: Research on prior knowledge and conceptual competencies show that encouraging a focus on them is important for mathematics instruction. A number of teaching techniques can be used to foster students' conceptual understanding of problems and elicitation of prior knowledge. Sweller, Mawer, & Ward (1983) and Siegler (1995) suggested the following: First, when possible, the teacher should try to illustrate the problem using contexts that are familiar and meaningful to students. In addition to fostering the students' conceptual understanding, familiar contexts will help students remember what has been presented in class. Second, after students have developed some skills in solving a type of problem, the teacher should present a few problems that are good examples of the type of problem being covered and have students solve the problems in a variety of ways. Psychological studies have shown that solving a few problems in many ways is much more effective in fostering conceptual understanding of the problem type than solving multiple problems in the same way. Solving problems in different ways can be done either as individual assignments or by the class as a whole. In the latter case, different students or groups of students might present suggestions for the problem. The students should be encouraged to explain why different methods work and to identify some of the similarities and differences among them. Third, for some problems, the teacher might have to explicitly teach one approach and then challenge the class to think of another way of solving the problem. The teacher could have students explain why the teacher used a certain strategy to solve a problem. Having students explain what someone else was thinking when he or she solved the problem facilitates their conceptual understanding.

Misconception 34

Exponents Addition

Grades: 7 to 9

Question: Simplify

$$Y^4 + Y^4$$

Likely Student Answer: $Y^4 + Y^4 = Y^8$

Explanation of Misconception: This is another misapplication and overgeneralization of rules. The student thinks that it is okay to add the powers because the base is the same for both terms. Instead of adding both powers, the correct thing to do would have been to add the coefficients of the two terms to obtain $2Y^4$.

What Teachers Can Do: The teacher should attempt to distinguish between $Y^4 + Y^4$ and $Y^4 \cdot Y^4$ analytically. Such distinction could reveal the student error. The use of a graphic organizer to illustrate the concept is suggested.

Apprise students that certain mathematical concepts are better learned using graphic organizers and this problem is one of such. The graphic illustration in Figure 34.1 shows that Y^4 factors out to be $Y \cdot Y \cdot Y \cdot Y$ and they see that Y^4 is actually $1Y^4$. The addition of $1Y^4$ and $1Y^4$ gives $2Y^4$. By way of distinguishing, it would also be beneficial to represent the $Y^4 \cdot Y^4$ in graphic form. They would see how that expression equals Y^8 and therefore different from $Y^4 + Y^4$. Apart from allowing students see relationships between ideas, the strategy can be useful in other situations such as helping students organize information. The use of graphic organizers takes away the confusion associated with exponents. For example, as seen in Misconception 17, students confuse the powers of exponents by thinking that 5^3 is 5 x 3 which equal15. But if this concept is represented with the aid of a graphic organizer,

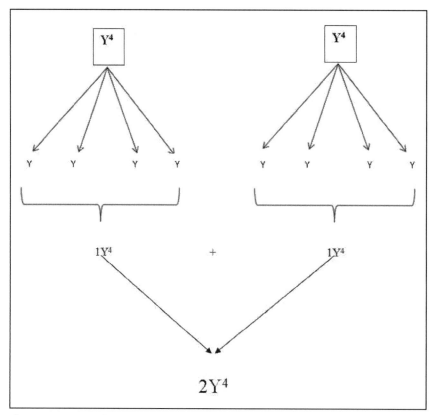

Figure 34.1. Graphic Organization of a Concept.

students would see how 5^3 is different from 5 x 3 diagrammatically. Teachers can use graphic organizers to reinforce learning, assess learning at multiple checkpoints, and identify misunderstanding of concepts. Graphic organizers can be used before, during, and after instruction. Teachers could use graphic organizers to brainstorm ideas, to activate prior knowledge, and to review. They are also valuable tools in any activity which requires the use of critical thinking.

Graphic organizers appear to be beneficial as instructional strategy that aid students in retaining learned information. Many students are visual learners, thus, a visual approach to brainstorming or organizing information is essential. As learners become familiar with using graphic organizers, they will develop their own approaches and create their own organizers. Encourage them to adapt them, change them, and create their own for more complex strategies and connections. Remember that there is no one right way to use graphic organizers; the best way is the way that works for each student.

Research Note: The literature supports the use of graphic organizers in facilitating and improving learning outcomes for a wide range of learners not only in mathematics but in other school subjects. Horton, Lovitt, & Bergerud (1990) reported on the value of graphic organizers to both middle school and high students with or without disabilities as an organizational tool to promote the memory of content-related information. Other research (e.g., Jitendra, 2002) indicates that organizers assists these same students in how to represent problem situations, such as searching for solutions to word problems in mathematics. Frequently, students with learning disabilities have difficulties recalling key information, making connections between broad concepts and detail, and solving mathematical word problems. According to Maccini & Ruhl (2000), students with learning disabilities might experience fluency difficulties with mathematical facts and with basic mathematical procedures. Teachers must be made aware that the use of graphic organizers is not only a valid instructional practice but a viable strategy that might lesson the difficulties that students with learning disabilities experience in mathematics (Gagnon & Maccini, 2000).

Ausubel (1963) believed that the manner in which knowledge is presented can influence learning. The appropriate organizer can help students form relationships between previously acquired knowledge and new concepts. Research shows that graphic organizers are key to assisting students improve academic performance. For example, Okebukola (1992) noted that data obtained in his study provide supportive evidence to indicate that the subjects in the study who were adjudged to be good concept mappers exhibited superior performance in solving the three problems of the study. Also, Willerman & Mac Harc (1991) reported the relevance of graphic organizers in science. A control group of 40 eighth graders completed a unit on elements and compounds. An experimental group of 42 completed *concept maps* on same topic. Results of a one-tailed t-test demonstrated the usefulness of *concept maps* as graphic organizers. These results are consistent with the findings of Sneed & Snead (2004) who suggested that lower achieving students appear to have success with the usage of concept mapping (advanced graphic organizers) in science.

Researchers (e.g., Ausubel and others mentioned here) have noted that graphic organizers aid comprehension for several reasons: Graphic organizers match the mind and because it arranges information in a visual pattern that complements the framework of the mind, it make possible for information to be easily learned and understood; organizers demonstrate how concepts are linked to prior knowledge to aid comprehension; organizers aid the memory as opposed to recalling key points from an extended test; organizers help retain information readily when higher thought processes are involved; organizers engage the learner with a combination of the spoken word with printed text and diagrams. Lenz et al (2004) pointed out the significance of graphic organizers as related to students' ownership of the learning process. According to the researchers, creating a graphic organizer for an instructional lesson plan is an effective way to engage students in learning and it also provides a way to integrate an additional learning modality into instruction.

Misconception 35

Zero Exponents

Grades: 7 to 9

Question: Simplify
   -5^0

Likely Student Answer: $-5^0 = 1$

Explanation of Misconception: The mathematical axiom (truism) that any number raised to the power of zero gives an answer of 1 must have necessitated this misconception. The student did not realize that the base for the problem is negative 5 and not 5. Thus in simplifying the problem there is the need to recognize that the power zero affects the 5 only and therefore the negative sign ought to appear in the answer: -1.

What Teachers Can Do: With this problem, it is important for the teacher to distinguish between -5^0 and 5^0. It should be pointed out that because there is no parenthesis encompassing the negative sign and the 5, it means that the sign (-) and the number (5) are separate entities. Therefore, the power of zero affects only the number 5. Also, a teacher should not stop at just stating the rules of exponents and thereafter make students apply these rules. The teaching should be extended to include negative numbers such as -5; expressions involving parenthesis; and fractional and negative exponents.

When teaching the zero exponent concept, it is suggested that the teacher use real-world approach before resorting to symbolic and analytic approaches. For example, a teacher could provide students with a sheet of paper and have them fold it in halves as many times as possible. They will probably achieve six or even seven folds. Since the number of rectangles (folded layers) doubles with every fold, they are growing exponentially. Exponents could be used to study the process. Let 2^n represent n doublings or folds of

No. of Folds	No. of Rectangles
1	$2 = 2^1$
2	$4 = 2^2$
3	$8 = 2^3$
4	$16 = 2^4$
5	$32 = 2^5$
6	$64 = 2^6$
7	$128 = 2^7$

Figure 35.1. Results of Paper Folding.

No. of Folds	No. of Rectangles
0	$\mathbf{1 = 2^0}$
1	$2 = 2^1$
2	$4 = 2^2$
3	$8 = 2^3$
4	$16 = 2^4$
5	$32 = 2^5$
6	$64 = 2^6$
7	$128 = 2^7$

Figure 35.2. Results of Paper Folding.

the paper. The table represented in Figure 35.1 shows what is happening to the number of rectangles as the paper is folded.

The right side of the table shows the total number of resulting rectangles. The exponent tells how many total folds have been made. With this understanding, what is meant by the following notation: 2^0? This has to mean that the paper has not been folded (not been doubled). But how many rectangles are there? Clearly, $2^0 = 1$ as shown in Figure 35.2.

Similar arguments can be made regarding tripling and quadrupling folds, which could be an extension to this activity. No matter what type of fold is made, the activity always begins with one layer of paper (i.e., one rectangular paper). Thus $n^0 = 1$ for all nonzero values of n. The pattern approach could be introduced at this juncture and note that it is similar to what we have on the tables. The approach is useful when dealing with negative exponents. For example, $2^{-1} = 1/2$, $2^{-2} = 1/4$, $2^{-3} = 1/8$, and so forth. After students have grasped the concept through non-traditional approaches such as discussed above, the teacher could then proceed to use symbols and analytic means in the instructional sequence. The teacher could use the laws of exponents and the properties of algebra to show.

Research Note: Some research findings have stated that starting the teaching of exponents with laws does not help students. For example, Lochhead (1991) stated that the standard introduction of exponents through laws may not be the best way to achieve strong conceptual understanding and may even contribute to superficial procedural concept of what exponents are and how the various definitions connect together to one consistent and powerful notion. Other research work has focused on knowledge and understanding. For example, Ojose (2012) investigated the content knowledge of teachers and their corresponding pedagogical knowledge of the zero exponent axioms. Findings of the study indicate that 50% of algebra teachers do not possess the content knowledge relating to the axioms. This lack of adequate knowledge manifested in ineffective instruction. For example, the teachers without knowledge of the axioms assumed that when they merely state rules, students get them and therefore the need for further explanation or activities to enhance the understanding of the axioms are not necessary. This wrong assumption leads to students being denied access to conceptual understanding.

Elstak (2007) examined the understanding pre-service teachers have of the concept of zero exponents. Pre-interviews with novice and expert mathematics students suggested that novice students had fragmented notions of exponents. In applications with all kinds of exponents (zero, rational, and negative), novice students relied on operational procedures and authority of teachers. Neither novice students nor expert students proposed integrated concept of exponents to explain all types of exponents. A conjecture for transforming the teaching and learning of exponents was proposed, that the teaching and learning of exponents can be improved through the study of the concepts of rate of growth and powers of factors. The conjecture was tested in a teaching experiment with the novice students. The role of the laws of exponents in the formation of rational, negative, and zero exponents was examined. The students' construction of the concept of the various kinds of exponents was described through models. The result suggests that students do not base their understanding of exponents on the patterns of the laws of exponents.

Misconception 36

Solving Equation by Addition and Subtraction

Grades: 5 to 8

Question: Solve
$$X - 7 = -7$$

Likely Student Answer: $X = -14$

Explanation of Misconception: This is a structural error. The student is unable to apply conventions of manipulating equations necessitated by lack of conceptual understanding. The student simply isolated the variable X by erroneously adding -7 and -7 to get -14. The student felt it right to subtract 7 (rather than add 7) to both sides of the equation.

What Teachers Can Do: This is one of the many problems that manifest as students transition from arithmetic to algebra because of their narrow understanding of the concept of "variables." The above highlighted misconception is only one of the various types that students will exhibit in equations solving, especially in grades 5 to 7. For example, some students believe that $5X + 7 = 20$ and $5Y + 7 = 20$ will have different solutions. Teachers should not rely on traditional didactic methods of showing and telling to teach these concepts. Teacher created activities, models, and even games are more helpful for students in learning the concept of variables and equations. For example, the following is an activity that might be helpful: Ask students how they know that $12 + 27 = 27 + 12$ without doing the computation. Students' explanation should show evidence of commutative property of addition. How can this be written to show it's true for every number, including negatives, and even fractions and decimals? If students do not suggest it, offer the idea that letters or shapes could be used like this: $\lozenge + \Delta = \Delta + \lozenge$ or $x + y = y + x$. Be sure students understand that the choice of letter or shape is totally arbitrary as long as it is understood that each stands for any number and that when the same letter or shape appears in the same equation, it must represent the same value. With this as the background, make students find other statements that are true for other numbers.

Also, translating story problems is helpful at this stage of transition to algebra. Read a simple story problem to students. Their task is to write an equation that means the same thing: *There are seven red pens on a table, how many more red pens do we add if we want the total number of pens on the table to be unchanged?* ($7 + \Delta = 7$). In other words, what goes in the place of the Δ? This activity can be reversed by providing an equation with unknown and having students make up a story to go with it. Once equations are agreed on, students should use whatever means they wish to find values that make the sentence true. Trial and error is a reasonable first strategy. Model such as balance could also be introduced at this stage and students can build on it as strategy for solving equations. For example, implore students to think of how many more red pens would be put on the other side of a weight balance with the above stated problem to make it *balance*. Physically manipulating the balance using the *pens* is quite ideal.

Games are a fun way of reinforcing the concept. One such game is called the equate game developed by Mary Kay Beavers for Conceptual Math Media, Inc. It is a game where players score points by forming equations on a 19 x 19 game board. The game is helpful to students because they learn and develop the concept of equality with it. The object of the game is to compel the players or teams compute and think strategically, critically, and creatively. In this game a player forms true equality statements that appear across and down in a crossword fashion and must be mathematically correct. It is similar to Scrabble game except that players use digits and mathematical operators instead of letters. After beginning at the center of the board, each successive play connects with a previous play. Players strive for a high score by trying to take advantage of both the individual symbol scores as well as the premium board positions. The game is recommended for reinforcing the concept of equations.

Research Note: The literature suggests that the use of models is beneficial to students. Modeling might be used both as a teaching and assessment tool since mathematical models might be viewed as external indicators of student cognitive structures that are built and amplified through the teachers' interventions. The most important goal of teaching mathematics is to instill a value of the possibilities of using mathematical methods to handle incoming problems from all different parts of life (Duncan et al., 1996). A number of studies have shown that the use of visual and concrete representations improves performance of students. In fact Simon (1981) argued that solving a problem simply means representing it so the solution is transparent. Simon further noted that learners are given ready models specially created for certain problems or situations and are required to change certain parameters in the model to be able to solve related problems. Learners are then expected to see the relationships between objects in the model and expected to construct mathematical concepts through mathematical abstraction.

The literature on the effect of games on teaching and learning of mathematical concepts alludes to their usefulness. For example, Whyte & Bull (2008) examined the effects of linear number board game play on basic number skills, including counting, number naming, magnitude understanding, and number line estimating. Groups of children played linear or nonlinear number or linear color games. The children in both number game groups showed significant improvements in counting abilities and magnitude understanding, but children in the linear color group showed stable performance between pretests and posttests. Results also indicate that playing linear number board games improved children's numerical estimation and comprehension. The researchers concluded that experience with number-based board games could drastically improve children's accuracy of number line estimates and the linearity of their numerical magnitude representations.

Ke (2008) studied the use of drill and computer games on cognitive achievement, metacognitive awareness, and attitudes toward mathematics learning. Data obtained from observation, analyses of computer-generated records, and think-aloud protocols showed that games reinforced mathematics standards. Ke's findings indicate that students developed significantly more positive attitudes toward mathematics learning through games. Ke also found that cooperatives, as opposed to a competitive individualistic goal structure, enhanced the positive effects of a game on attitudes toward mathematics learning. Results also showed that peer communication was mostly play-based rather than learning-oriented, and computer gaming had a nonsignificant effect on mathematics test performance. Ke concluded that suitable off-computer activities, such as offline assistive learning tools, game-based collaborative activities, and the just-in-time guidance of an instructor, enhance the game-based learning process. Burns (2009) provided the following tips for math games: Choose a game that are accessible to all students; play cooperatively and competitively; choose games that require reasoning and chance; teach the game to the entire class at the same time; and start a math games chart that students could easily see.

Misconception 37

Solving Equation by Division and Multiplication

Grades: 5 to 8

Question: Solve
$$180x = 2$$

Likely Student Answer: $x = 90$

Explanation of Misconception: The student is applying certain mathematical rules to the wrong situation. The student probably thought that because 2 multiplied by 90 gives 180, then 90 ought to represent the value of x. Instead of thinking of the number that could multiply 180 to produce 2, the student erroneously dwelt on finding a number that could multiply 2 to produce 180. This is an arbitrary error related to conceptual understanding.

What Teachers Can Do: In addition to some of the strategies suggested in Misconception 36, teachers should provide opportunity for students to check the reasonableness of their answers. For example, quick answer check strategy will reveal to students that 90 as an answer to the above problem is not reasonable. If possible have them use calculation device like a computer or a calculator. Better still, the teacher could state the problem in words for students and let them ponder it: *What number will you multiply by 180 to get 2?* Again, this will make students think about the reasonableness of the answer. A teacher should first of all give students the opportunity to explain their answers. In a situation where a student reasoning is incorrect, the teacher should guide students to the right thinking. Also, teachers should insist on the use of mental math as this would also make them think about the reasonableness of an answer. The use of mental math should deliberately be included in a teacher's repertoire. If mental math is constantly practiced in the classroom, students are better equipped in using the strategy when solving mathematics problems.

There are also software programs available online that teachers could use when teaching equations. Even though they do not substitute for good instruction, teachers should use them when appropriate. For example, eigenmath is an online program that can solve this and other kinds of equations for students or give them feedback by highlighting what they did wrong. There are many of such programs online and are free for users (e.g., Apple or Google Apps). Another online tool for learning mathematics concepts like equations is virtual manipulatives. With them, students can manipulate all kinds of objects (algebra tile, cubes, apples, oranges, etc) in picture form and obtain same results as they would with physical objects.

Research Note: Two research areas related to this topic as highlighted above are mental math and virtual manipulatives. Mental math are quite useful in handling one step equations of this type as they do not require elaborate solving procedures. The literature highlights the importance of doing mental math when appropriate. Some of the research focuses on the nature of mental math and estimation as children navigate through the grades. For example, Siegler and Booth (2004) did work on the development of numerical estimation in young children. Their initial analyses examined the fit of linear and logarithmic functions to the estimates of children, grouped by grade level, for number line estimate items. Results showed that, for grades K-2, estimates became more linear with increasing age and experience and the slopes of the best fitting linear functions moved increasingly toward 1.0, the ideal slope relating estimates to the numbers presented in the test items. Results also showed the same developmental sequence for grades 2 – 6. Results supported providing more opportunities for children to locate a wide range of numbers on number lines and for encouraging children to locate benchmarks on number lines and using these benchmarks to locate other numerical magnitudes. Laski and Siegler (2007) followed up the earlier numerical estimation study by investigating connections between children's performance on tasks to categorize, compare, and estimate numbers. Their findings led to the suggestion that teachers provide children with experiences to encourage the use of linear representations over wider numerical ranges.

With virtual manipulatives, the literature supports their use but also offer some words of caution. A virtual manipulative is defined as "an interactive web-based visual representation of a dynamic object that presents opportunities for constructing mathematical knowledge" (Moyer et al., 2002, p. 373). Visual representations of concepts and relations helps learners gain insights in mathematics. Virtual manipulatives enable as much engagement as physical manipulative do since they are actual models of physical manipulatives that include tangrams and geoboards (Dorward & Heal, 1999). They may provide interactive environments where students could pose and solve their own problems to form connections between mathematical concepts and operations, and get immediate feedback about their actions that might lead them to reflect on their conceptualization. Virtual manipulatives *must* be designed in a way to put focus on the mathematical concepts conveyed making their functionality as transparent as possible.

Ozmantar (2005) noted that computer manipulatives could be used to reinforce conceptual understanding. They could also be used to design extra-curricular activities since they are easily accessible both at home and the schools. Clements & McMillen (1996) offered the following recommendations for selecting and using virtual manipulatives in a learning environment: (a) Use computer manipulatives for assessment as mirrors of students' thinking; (b) Guide students to alter and reflect on their actions, always predicting and explaining; (c) Create tasks that cause students to see conflicts or gaps in their thinking; (d) Have students work cooperatively in pairs; (e) If possible, use one computer and a large – screen display to focus and extend follow-up discussion with the class; (f) Recognize that much information may have to be introduced before moving to work on computers, including the purpose of the software, ways to operate the hardware and software, mathematics content and problem solving strategies; and (g) Use extensible programs for long periods across topics when possible.

Misconception 38

Fractional Equations

Grades: 6 to 9

Question: Solve
$x/4 = -4$

Likely Student Answer: $x = -1$

Explanation of Misconception: This is also a misconception associated with misapplication of mathematical rules. To the student, simply dividing -4 by 4 to get -1 suffices as an approach to isolating variable X for the problem. The knowledge of when and what to divide or multiply when isolating a variable is lacking. Multiplying both sides of the equation by 4 would have isolated the variable X and thus produced the right answer.

What Teachers Can Do: Three strategies for teaching this concept are suggested: State problem in words, teach multiple analytic methods, and use a table to find the unknown. First, if this problem is stated in words, it may be easier for students to relate with. For example, a teacher could state the problem the following way: *What number will 4 divide to arrive at -4?* When stated in words, the teacher should encourage students to come up with answers that they think are reasonable. The class should then discuss all responses while the teacher moderates and help students correct their errors. Secondly, based on prior knowledge of students' equation solving, multiple analytic methods can be introduced by the teacher. For example, the teacher could explain a method of clearing the fractions and solving linearly. Another method is to cross multiply the fractions. Either of these two analytic methods requires students to have prior knowledge of simple algebraic manipulations. The drawback however is that, if students are not ready because they lack the background knowledge required of the problem, the situation is further com-

X	X/4	Answer
-20	-20/4	-5
-16	-16/4	-4
-8	-8/4	-2
-4	-4/4	-1
-2	-2/4	-1/2
0	0/4	0
2	2/4	1/2
4	4/4	1
8	8/4	2

Figure 38.1. Table of Values.

pounded. It is therefore suggested that the teacher diagnose students' prior knowledge before using the above referred analytic methods. Third, the use of a table to organize the learning is suggested.

The teacher should suggest various values of X to substitute into the expression x/4 and have students organize the result in a table as shown in Figure 38.1. The goal is to find an X value that would make x/4 simplify to -4. The students will discover that substitution of X values into x/4 will produce only one result that equals -4 and that value is -16. As highlighted, when -16 is substituted in x/4, the result is -4 and therefore the answer to the equation. This is the same answer that would be arrived at if solved algebraically. Since this is more or less a trial and error approach, it is important to suggest to students that they do not skip numbers like is done in Figure 38.1. In other words, when students are using this approach, every number in the range should be substituted because, a skipped number might just happen to be what they are trying to find. Apprise students that all equation problems which usually require finding an unknown can be figured out by the process of substituting numerical values into the equation sentence to find a number that makes the sentence true. Tables can generally be used to facilitate the process.

Research Note: Swafford & Langrall (2000) investigated students' intuitive knowledge of algebra prior to formal instruction. More specifically, they investigated sixth-grade students' ability to solve problems involving specific cases, to generalize problem situations using verbal descriptions and equations, and to solve related problems using their equations. They showed that although prior to studying algebra students may not view equations as mathematical objects in their own right, they do not begin their formal study of algebra without prior knowledge. They stated that there is potential benefit in examining the same problem through different representations such as diagrams, graphs, tables, verbal descriptions, and equations. However, the use of multiple representations in and of themselves is not enough. For example, in the case of tables, students could identify isolated patterns between pairs of dependent and independent variables but could not see a pattern that was consistent across the entire table. As Lee (1996) noted, the problem is not the inability to see pattern but the inability to see an algebraically useful pattern. Tables were most useful when they were produced by students trying to make sense of the problem context instead of when provided or suggested. Whatever the representation, students need to make the link between the representation and the problem context and between one representation and another.

Dreyfus (1991) has suggested that the learning process needs to proceed through four stages: (a) using a single representation, (b) using more than one representation, (c) making links between parallel representations, and (d) integrating and flexibly switching among representations. Dreyfus noted that instruction is too often focused on the first two stages, with the assumption being made that students will reach the other two stages as a by-product of activities in the earlier stages. At the prealgebra level, the emphasis in the curriculum should be on developing and linking multiple representations to generalize problem situations instead of on merely constructing representations that students do not link to problem situations.

Misconception 39

One-Step Inequality

Grades: 7 to 12

Question: Solve
 $-5X < 10$

Likely Student Answer: $X < -2$

Explanation of Misconception: Most students would apply the "normal" rules of solving equations even when they are dealing with inequalities. In this case, the student rightly divided both sides by -5 but did not pay attention to the inequality symbol. The student did not reverse the inequality symbol to make the inequality statement true. This is an arbitrary error associated with lack of conceptual understanding.

What Teachers Can Do: Thorough understanding of equivalence and equality concepts are key to subsequent learning of inequalities. As has been highlighted elsewhere in the book, students lack understanding of these concepts in part because, instructional strategies do not emphasize relational meaning. That (mis)understanding is compounded when the "<" and ">" signs are introduced to students. Further compounding the issues is the fact certain acceptable ways of writing the signs are not conventional. For example, $a < b$ can be written as $b > a$. Teachers should therefore form the habit of writing answers in various ways with equations (e.g., an answer with $x = 4$ can alternatively be written as $4 = x$) in order to mitigate these problems.

Teachers should show the relationship that exists with equality, equal signs, equations, and inequalities. Teacher – created activities could be helpful in this regard. For example, a teacher could challenge students to find different ways to express a particular number, say 10. Give a few simple examples, such $5 + 5$ or $12 - 2$. Encourage students to use two or more different operations. Also, on the board, a teacher could draw a simple two-pan balance. In each pan, write a numerical expression and ask which pan will go down or whether the two will balance. Challenge students to write expressions for each side of the scale to make it balance. Point out that when the scale tilts, either a greater than (>) symbol or a less than (<) symbol is used. Explain why it is important to reverse the direction of the inequality symbol when dividing or multiplying by a negative number. For example, the inequality 2 < 4 is true. If you multiply both sides by -1, the result without reversing the sign is false: -2 < -4. The teacher should emphasize that when dividing or multiplying an inequality by a negative number, one must always reverse the direction of the inequality symbol so that the new inequality is equivalent to the original.

Certain strategies could help students understand why the direction of the inequality symbol must be reversed when dividing or multiplying by a negative number. One is to have them divide several numerical inequalities by negative numbers. They will discover that the inequality is false after division unless the sign is reversed to maintain equivalence. Two is to have students hypothesize whether they must change the direction of the inequality symbol when they multiply or divide by a negative number. Have students perform the operation to test their hypothesis. Lastly, have students write rules for dividing or multiplying inequalities on a note card. Make sure students include examples of how to change the direction of the inequality on their note cards. Suggest to students that they refer to their note cards when solving inequality problems.

Research Note: Rittle-Johnson & Alibali (1999) studied the relationship between children's conceptual understanding of mathematical equivalence and their procedures for solving equivalence problems and found causal relationship between conceptual and procedural knowledge that suggests that conceptual knowledge may have a greater influence on procedural knowledge than the reverse. Although the evidence presented by their data on equivalence suggests an iterative relationship but the influence of one on another may not be symmetrical. Children who received procedural instruction showed greater gains on both the question and evaluation tasks than children in the control group, but these gains were modest and were smaller than the gains made by children who received conceptual instruction. Children in the procedural-instruction group also had significantly poorer transfer performance than children in the conceptual group. Their data provides some clue about possible mechanisms of the influence of both kinds of knowledge on one another. On how conceptual knowledge influence procedure generation, one possibility offered by the data points to the fact that gains in conceptual knowledge helped children to recognize that their problem-solving procedures are incorrect. As children solidify their conceptual knowledge, they may begin to realize that their problem-solving procedures are inconsistent with that knowledge. By highlighting such inconsistencies, gains in conceptual understanding may lead children to change their problem-solving procedures. Thus, conceptual knowledge may influence procedure generation by helping children recognize the need for new procedures.

Two possibilities were offered on how procedural knowledge might influence conceptual understanding. One possibility is that procedural knowledge constrains conceptual understanding. Knowledge of correct procedures may help children to zero in on accurate conceptual knowledge. For example, correct procedures for solving math equivalence problems are inconsistent with naïve conceptions of the equal sign as meaning "get an answer" and as signaling the end of the problem. In the study, procedural instruction led to improved understanding of the equal sign and its implications for the structure of equations. Thus, procedural knowledge could influence conceptual understanding by helping to eliminate misconceptions. A second possibility is that, when children use procedures, they sometimes think about why they work. When procedures are easy to implement, children may not use all their resources in implementing the procedure, and they may have resources available to consider the basis of the procedure. Further, in situations in which newly learned procedures result in different solutions than prior procedures, children may find this surprising and this may lead them to consider the conceptual basis of the new procedure.

The findings of Rittle-Johnson & Alibali are consistent with findings of other researchers as it pertains to inconsistencies in procedural and conceptual knowledge. For example, while Perry (1991) found that instruction on the concept of mathematical equivalence led to substantial minority of forth-and fifth grade students to generate a correct procedure for solving equivalence problems, Sielger (1991) showed that correct procedure aids the conceptual understanding as seen in their study of children counting correctly before they understand certain counting principles, such as the irrelevance of counting order. Overall, the literature suggests that conceptual understanding plays an important role in procedure adoption and generation. However, it seems likely this relationship is not unidirectional one. Instead, conceptual and procedural knowledge may develop iteratively, with gains in one leading to gains in the other, which in turn trigger new gains in the first.

Misconception 40

Absolute Value

Grades: 7 to 9

Question: Insert <, >, or = in the number sentence.

$$|-3| \;\square\; |3|$$

Likely Student Answer: $|-3| < |3|$

Explanation of Misconception: The student does not possess a conceptual knowledge of absolute values. The student is not recognizing that $|-3|$ is the distance between -3 and zero on the number line which comes out to be 3 units. The same is true for $|3|$ which is also 3 units from the point zero on the number line.

What Teachers Can Do: This misconception could persist even after an instruction that is not hands-on based is delivered to students. In other words, teaching this concept with rules may not be beneficial to students. The teaching should involve models such as the number line. With the number line, students see that the absolute value of an integer is the distance between the position of the integer and zero on the line:

In Figure 40.1, we can see that the distance between zero and 3 and the distance between zero and -3 are the same; hence absolute value of 3 for both distances. The teacher could challenge students to come up with a general rule for absolute value of numbers based on the number line understanding. It is more beneficial for students to conceptually understand the idea of absolute values and then come up with rules instead of being handed the rules and make a lesson revolve around those rules.

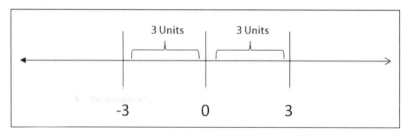

Figure 40.1. Representation of Absolute Values.

Another approach that could be used to address this kind of concept is called the "sucker bet" approach in which students look at a mathematics problem and try to imagine what their instinctive answer might be. The teacher query is then posed as a question–"what do you think some people might say?"–in order to de-personalize the process, making it safe for students to answer. Students then answer with reasons why people might think their answers were correct. Following a discussion of possible sucker bet answers, the teacher models the possible answers and leads students to disprove the sucker bet answers and prove the correct algorithmic approach. Following this, the students are asked to demonstrate their understanding by making up similar sucker bet problems for their peers. For this concept, the sucker bet could be: "Of <, >, and =, which do you think some people might write as the symbol satisfying $|-3|\ \square\ |3|$?" Why would they write that particular symbol? Having students explore what their gut reaction answer is to a problem, clarifying the correct answer, and creating and discussing similar problems can uncover misunderstandings and misconceptions. This approach also lends credence to open-ended questioning technique that is quite valuable in teaching and learning of mathematics.

Research Note: Students develop an understanding of mathematics by linking new ideas to an existing schema. Even though students may appear to understand more advanced concepts, they will often make mistakes and revert to old error patterns when presented with problems in a new context. The sucker bet approach has been written about by Johnson (2008) in which she noted that the approach could help students break the cycle of misconception in that it helps students examine their instinctive reaction to a problem. According to Johnson, teaching for understanding must begin with three basic steps: Eliciting from the child what his/her understanding about a concept is, remediating any misconceptions, and using active learning to build conceptual understanding. Johnson claims that this process is enhanced with an approach such as the "sucker bet" approach. According to Johnson, when children search for meaning in mathematics and or try to keep the rules for different processes straight in their heads, they may revert to their incorrect algorithm that could yield the sucker bet.

With regard to open-ended response and open ended questioning, researchers agree that teachers should form a habit of using them in facilitating instructional procedures. Millhiser (2011) suggested four benefits of open-ended and open-response questioning in teaching: (a) Open-ended questions helps students build confidence by problem solving in ways that come naturally; (b) An open-ended question, like public speaking and journal writing, is especially helpful in developing the students' writing ability to communicate; (c) Such questions can communicate levels of student achievement more clearly and gives better guidance for instruction; and (d) Open- response questioning promote acquisition in conceptual knowledge, unlike procedural knowledge, is more readily applied in unfamiliar contexts outside the classroom.

Misconception 41

Operations with Radical Expressions

Grades: 8 to 12

Question: Simplify

(a) $\sqrt{(9 + 16)} = ?$

(b) $\sqrt{24} - \sqrt{6} = ?$

Likely Student Answer:

(a) $3 + 4 = 7$

(b) $\sqrt{18}$

Explanation of Misconceptions: (a) The student is misapplying the knowledge of finding square root of numbers. The student ought to decipher when it is necessary to work out an operation in the radical sign before finding the square root. Instead of adding up the terms and finding the square of the numbers, the student erroneously worked out the individual square root of both numbers as if the operation symbol was multiplication. (b) The correct knowledge of radicals requires students to differentiate between radicands with perfect squares and non-perfect square radicands. In the expression, the student erroneously subtracted the radicands as if he/she is dealing with natural counting numbers. The correct thing to do would have been to find the factors of 24 (4 and 6), of which the 4 is a perfect square. Therefore $\sqrt{24}$ could be expressed as $2\sqrt{6}$. Because $2\sqrt{6}$ and $\sqrt{6}$ are alike, the subtraction operation could be performed.

What Teachers Can Do: The ambiguity of the radical sign which is necessitated by change from its arithmetic meaning to the algebraic meaning often goes unnoticed by teachers and textbooks. The ambiguity may be cause of certain cognitive conflicts for students. To help avoid the conflicts and misconceptions, teachers should plan instruction that scaffolds student's learning that move from very basic and simple to high level and complex skills. For example, a teacher should acquaint students with the dual property of the radical sign: $\sqrt{4} = \pm 2$. This dual nature could be explained with the product property of numbers: 2 times 2 = -2 times -2 = 4. The concept of perfect squares and how they relate to the radical sign should also be demonstrated. The teacher could start with single digit perfect squares and ensure understanding before moving to double digits perfect squares.

Next, instruction should move to conjugate radicals like $\sqrt{4} + \sqrt{4}$. It is important to emphasize that $\sqrt{4} + \sqrt{4}$ is quite different from $\sqrt{4+4}$. While the former gives 2 + 2, the latter does not. The simplified form of $\sqrt{4+4}$ is $\sqrt{8}$. Next, the teacher should move to introduce different combinations like $\sqrt{9}$ and $\sqrt{16}$. Point out that with same reason as highlighted above, $\sqrt{9} + \sqrt{16}$ and $\sqrt{9+16}$ are quite different. While $\sqrt{9} + \sqrt{16}$ simplifies to 7, $\sqrt{9+16}$ simplifies to 5. Make students discuss the differences and let them come up with strategies for recognizing such differences when they are given radicals to simplify. One strategy that could make students familiarize themselves with radicals is to write various radical expressions for the same number. For example, a teacher could give a number, say 2, and have students come up with several radical expressions that simplifies to that number. Some expressions that students might come up with may include: $\sqrt{4}$, $2 \cdot \sqrt{1}$, $\sqrt{16} \div \sqrt{4}$, and so forth. Some students would in all likelihood give $\sqrt{1} + \sqrt{1}$ as an expression that simplifies to 2. A teacher could capitalize on that relationship to further distinguish between $\sqrt{1} + \sqrt{1}$ and $\sqrt{1+1}$. Students could then apply the same reasoning to distinguish between $\sqrt{9} + \sqrt{16}$ and $\sqrt{9+16}$ and hence be able to explain why $\sqrt{9+16}$ does not simplify to 7 but 5. In this discourse, the teacher should focus on sense making of the radical sign as opposed to mere computations. Students should explain their reasoning and justify answers.

It is also important to debunk the notion held by some students that answers should be free of radical signs. Students should learn to decompose terms in a radical expression and extend the rules of combining like terms to simplifying the expressions. For example, with problem (b) above, $\sqrt{1}$ decomposes to $2\sqrt{6}$. Students can then perform simple subtraction operation because the two terms ($2\sqrt{6}$ and $\sqrt{6}$) are alike and therefore can be simplified.

Research Note: Much of the literature on radical signs relates to its ambiguity as a source of problem for students. With regard to cognition, Kieran (2006) noted the need to reconceptualize signs that change meaning when passing from arithmetic to algebra. This happens with the radical sign which changes, since it either indicates an operation, as happens in $\sqrt{4}$, or indicates the main root of this operation, as happens in the solution to the equation: $X^2 - A = 0 \rightarrow X = \pm\sqrt{A}$. According to Roach et al (2004), the duality of meaning causes contradictions because $\sqrt{4} = 2$ is acceptable in arithmetic but this soon changes in algebra since the square root of A (A < 0) cannot be calculated, so that to indicate its value the expression $\sqrt{A}$ is introduced, which no longer represents an indicated operation but result. The contradiction has manifested in ambiguity as in the case of simplifying $\sqrt{4} + \sqrt{4}$ which has multiple solutions: {-4, 0, and +4}. According to the researchers, this ambiguity does not help instruction.

Sfard (1991) pointed out the psychological aspects of the dual operational nature of mathematical conceptions and their role in the formation of concepts. Sfard noted that a mathematical entity can be seen as an object and a process. Treating a mathematical notation as an object leads to a type of conception called *structural*, whereas interpreting a notion as a process implies a conception called *operational*. For Sfard, the ability to see mathematical entity as an object and a process is indispensable for deep understanding in mathematics, such that the concept formation implies that certain mathematical notions should be regarded as fully developed only if they can be conceived both operationally and structurally. It is worth mentioning that when referring to the role of operational and structural conceptions, Sfard conjectures that when a person gets acquainted with a new mathematical notation, the operational conception is usually the first to develop, whereas the structural conception follows a long and difficult process that needs internal interventions (of a teacher, of a textbook), and may therefore be highly dependent on a kind of stimulus (of teaching method) which has been used.

Misconception 42

Simplifying Polynomials

Grades: 8 to 12

Question: Simplify
    (a) $6X^2 - X^2$
    (b) $(5X^2 + 4X - 3) - (2X^2 - X)$

Likely Student Answer: (a) 6
    (b)

$$\begin{array}{r} 5X^2 + 4X - 3 \\ -\ \underline{2X^2 - X} \\ 3X^2 + 3X - 3 \end{array}$$

Explanation of Misconceptions: (a) Instead of subtracting the coefficients of like terms which would produce $5X^2$, the student erroneously subtracted X^2 from X^2 leaving 6 as an answer. Clearly the student has no understanding of polynomials involving single terms, also known as monomials. This is a structural error related to conceptual understanding. (b) The misconception here is that the student added 4X and -X but actually needed to add 4X and the opposite of -X, since it is a subtraction problem. In other words, the middle term for the answer ought to be 5X and not 3X. This is an execution type of error because the student knows something but the implementation process breaks down somewhere along the line. This misconception is quite common with students in first algebra course, especially at the introductory level of polynomials.

What Teachers Can Do: With (a), it is important to highlight the differences in the two expressions so students could see that structurally, they are not the same: $6X^2 = 6 \cdot X \cdot X$; and $X^2 = X \cdot X$. Also, get students to "appreciate" the similarities between the expressions. A teacher could use a statement such as "We have six X^2 and one X^2 and subtracting them would only result in obtaining some X-squares." Rewrite the expression in a way that is familiar to the students:

$$6X^2$$
$$- \underline{1X^2}$$
$$5X^2$$

With this kind of representation, a teacher should emphasize that the two expressions are alike and can be operated on just like we do numbers. Remind students that the coefficients are like objects that are similar and can be added or subtracted. It is also important to relate these to some real-life contexts like money: *Six dollars minus one dollar*. With (b), one obvious issue is that the student is confused with signs. The teacher could use some kind of graphic organizer to illustrate the product or quotient of positives and negatives as shown in Figure 42.1.

Make students have this reference material and should be kept where they can access it when solving mathematics problems. The teacher may also encourage students to 'lock' these rules in their long term memory because they will encounter them throughout the study of arithmetic and algebra. In other words, encourage students to be sure they are using the right sign in any mathematics solving situation. They should not guess.

Another problem exhibited by this student is that of simplifying. The teacher should encourage students to avoid using a method that they are not

x or ÷	+	−
+	+	−
−	−	+

Figure 42.1. Rules for Multiplication and Division of Positives and Negatives.

comfortable with. For example, as shown above, this student used the column method of simplifying but was not good at it. The method led the student to disregard the effect of the minus sign on the subtrahend. An alternative method of working the problem vertically is suggested instead. This would mean rewriting the problem by opening up the parenthesis and then combining like terms. This would have left the student with the expression: $5X^2 + 4X - 3 - 2X^2 + X$. In this format, it is possible to combine like terms and write the answer. The use of algebra tiles is quite beneficial and is suggested in teaching the concepts.

Research Note: Matz (1980) reported on execution type of error. This is the type of error in which the student has some knowledge of the concept but not enough to complete the task. According to Matz, unlike planning errors that students do not have control over, execution errors are about the activity of an "executive" that control their application. Matz pointed out that the nature of algebra accounts for the "execution" errors. According to Matz, unlike arithmetic that has a single method of execution, algebra has "various features" which can often lead to breakdown during the process of execution. Matz suggested extrapolation (ways to project what one knows onto an unfamiliar situation) as a technique to remedy the execution problem. Matz cited "Linear Decomposition" as one form of extrapolation technique to deal with polynomials. This technique involves working with decomposable objects, whose parts can be treated independently.

Misconception 43

Systems of Equations

Grades: 8 to 12

Question: Solve the system of equations
$$4X - Y = 17$$
$$-X - Y = 7$$

Likely Student Answer:

$$4X - Y = 17$$
$$-\ X - Y = 7$$
$$3X\ \ \ \ = 24$$

$$X = 8, \text{ and } Y = 15$$

Explanation of Misconception: The student made an execution error as there seems to be some knowledge of what needs to be accomplished. The student knows that some kind of technique needs to be applied to get the values for X and Y. However, the procedure for obtaining the values breaks down because of misapplication of rules. It is not possible to eliminate the two negative Y's in both equations by addition. The right way would have been to subtract the second equation from the first one, giving a solution set of (2, -9).

What Teachers Can Do: Foremost, students should develop schema for variables. In other words, the schema for one-variable equations should be developed before embarking on systems of equations with students. Obviously, a lack of understanding with one variable equation could adversely

Systems of Equations

18	33
10	?

Figure 43.1. Sum-Up Activity: Finding Sum of Numbers.

affect learning systems of equations. This deficiency should be addressed by designing activities that help students acquire conceptual understanding. For example, to build up numerical strategies, a teacher could provide boxes as shown below and ask students to write answers in the spaces provided.

With Figure 43.1, students should provide an answer by adding up all three numbers; and with Figure 43.2, students should write different sets of numbers that could add up to the number provided. For example, 25, 10, and 20 would be one set of numbers for Figure 43.2. Encourage students to come up with as many sets as they can generate and provide opportunity for them to explain their choices of the numbers. The teacher should also extend the activity to include negative numbers. For example, -7, 70, and -8 would go in the boxes in Figure 43.2. With this activity, students could think flexibly and the process provides an avenue to generate mental strategies free from the vertical disposition of numbers. This and other strategies like differentiated statements of equality and writing answers in various forms could help students construct schema for variables. Again, the understanding of variable, equality and equivalence concepts are crucial to systems of equations.

One attribute of systems is that there are multiple ways of solving it: Elimination method, substitution method, linear combination method, graphical method, and of course a table. The teacher is encouraged to teach students all

?	?
?	55

Figure 43.2. Sum-Up Activity: Finding Numbers That Make Up a Sum.

methods and should encourage them to use one method to check another. In general, the technique of checking answers to see if an equation or systems of equations are solved correctly is very important. A substitution method could also be used to check if a solution set is correct. Students could, for example, be asked to substitute the solution set (8, 15) into the two original equations. They will discover that the set is not a solution. As a way of reinforcing the concept, a teacher may give multiple solution sets in which one set is right and the others are not. The students' task is to find the correct set and then explain why the others are not solutions. It is beneficial to put students in groups for the purpose of this activity so they can discuss their solutions. You might be amazed at how students can correct their own misconceptions through self-checking of solutions.

The graphical method is highly recommended in teaching systems as it opens itself up to real-life contexts. For example, cell phone problems are popular with systems. With knowledge of graphing, students are able to plot two graphs on the same plane and find the point of intersection of the graphs. The teacher can then extend that knowledge to finding out about two cell phone plans in which students would decide which plan is better and why. Two things can come from this activity: One is that students are able to read off the point of intersection of the two graphs as the solution set of a system. Two and more important, students could determine which cell phone plan is better at any given point in time. This is a higher level learning of the concept as it involves critical thinking skills.

Research Note: Trigueros et al (2007) acknowledged the difficulties associated with understanding the concept of systems of equations. They include the difficulty in deciding the number of solutions of a given system; presenting those solutions geometrically; interpreting graphs related to systems of linear equations; and passing from one form of representation to another. In order to study in- depth the nature of students' understanding and to identify sources of difficulties, these researchers designed an empirical study based on APOS theory (Action, Process, Object, and Schema) (Weller, et al., 2003).

In the study, six students who were taking a course based on APOS theory were interviewed at the beginning of the course and at the end of the course to study the viability of a proposed decomposed genetic decomposition, their difficulties, their reasoning pattern and the evolution of their schemas. They developed a possible genetic decomposition of solution to a linear system of equations and its possible evolution, which consists of the constructions the researcher thinks that students should make in order to construct the concept and those coordinations between them that underlie the evolution of the defined schema, and then analyzed activities that students can perform to make the necessary mental constructions. Schemas for equality and equations are coordinated to construct a process that transforms an equation into an equivalent one. The process is coordinated with the schema for system to construct a process to find an equivalent system of equations, and a process to determine the solution set from this equivalent form. This process of finding the solution set of a system of equations is encapsulated into an object, and then it is possible to study its properties and to relate it to its geometric interpretation.

Results of the study shows that the construction of a schema for variable that includes the interpretation and differentiation between the different uses of variable: unknown, general number and variable in functional relationship, and the understanding of solution of an equation as an object constitute necessary conditions for students to make the constructions that are needed in order to construct the systems of equation schema. The lack of those previous notions in the interviewed students seems to interfere severely with their possibility to make the constructions needed to understand new abstract concepts. In synopsis, the results indicate that an important prerequisite for students to profit from the opportunities of deepening students' understanding of linear systems of equations depends strongly on the development of their schema for variable, and that instructional strategy based on the APOS theory helps students in the evolution of their systems of equations schema.

Conclusion

No single method of instruction is the best or most appropriate in all situations. Teachers have a wide choice of instructional strategies for any given lesson. Teachers might use, for example, direct instruction, investigation, classroom discussion, drill, small groups, individualized formats, and hand-on materials. Good teachers look for a fit between the material to be taught and strategies to teach it. They ask: What am I trying to teach? What purposes are served by different strategies and techniques? Who are my students? What do they already know? Which instructional techniques will work to move them to the next level of understanding? Drawing on their experience and judgment, teachers should determine the balance of instructional strategies most likely to promote high students achievement given the mathematics to be taught and their students' needs. While embarking on this, care should be taken by the teacher to avoid over-reliance on computational and procedural skills to the detriment of conceptual understanding. This is important in a standards-based teaching in which good lessons are carefully developed and are designed to engage all members of the class in learning activities focused on student mastery of specific standards. Such lessons connect the standards to the basic question of why mathematical ideas are true and important. As a result, lessons will need to be designed so that students are constantly being exposed to new information while practicing skills and reinforcing their understanding of information introduced previously. The teaching of mathematics does not need to proceed in a strict linear order, requiring students to master each standard completely before being exposed to the next, but it should be carefully sequenced and organized to ensure that prerequisite skills form the foundation for more advance learning. In sequencing and organizing standards to maximize students learning, attention needs to be paid to issues that limit students learning of the subject; one of which is misconception.

Conclusion

While acknowledging the fact that misconceptions with students in K – 12 are many and can hardly be captured in its entirety in a single book, the assembly of common misconceptions in this book should be a starting point for teachers who genuinely want to help their students succeed in mathematics. As repeated several times in the book, misconceptions that teachers fail to mediate on gets much more difficult to address as the student progresses through the grades. The book has covered the common misconceptions in arithmetic and algebra. With each identified misconception, the error of students are highlighted and the possible reasons why students think the way they do is discussed. In *What Teachers Can Do*, strategies that teachers could use to teach the skill related to the concept are discussed. In some instances, the things that teachers should avoid are also discussed.

The strategies presented in the book are varied but most importantly, they are not in the direction of traditional methods that seems to be prevalent in most classrooms. These methods includes use of models; effective scaffolding; problem solving; use of graphic organizers; use of invented strategies; effective questioning; use of multiple representation (words, symbols, tables, graphs, etc); use of counter examples and counter arguments; use of games; use of nonroutine problems; teacher created activities that involves cutting and folding of papers; and so forth. Emphasis on traditional algorithms and didactic approaches to teaching are reasons for these misconceptions in the first place. Therefore, teaching in ways that differs from such tenets is a step in the right direction. The book has presented teaching strategies that are nontraditional but consistent with constructivist approaches. The teacher should employ strategies that they are comfortable with, but most importantly, such strategies must be useful in helping their students learn in meaningful ways so misconceptions are corrected and avoided.

The discussion of research related to these misconceptions shed light on many things: Sources of difficulties, strategies for instruction, students' thinking, and general promising practices. The *Research Note* explicates the views of experts with regard to each covered concept and misconception. As always with research, findings and pronouncements are not always in agreement when reporting on the same topic. The contradictory nature of research should help teachers ascertain that while some methods of teaching might work for certain groups of students and situations; the reverse might actually be the case for others.

Appendix A

List of Manipulatives and Their Uses

Table A.1.

Manipulative	Suggested Alternative	Use
Color Tile	Blocks, Buttons	Explore sorting, counting, and graphing
Attribute Blocks	Pasta, Buttons	Assists in logical thinking
Clock	Paper Plate, Brads	Beginning to end of a task is found by manipulating the clock
Connecting Cubes	Paper Clips	Constructing patterns
Two-Colored Counters	Buttons, Coins, Beans	Addition and subtraction of integers
Bucket Balance	Ruler, Paper Cups, Strings	Solving and determining unknown values of equations
Base-Ten Blocks/Models	Grid Paper	Composing and decomposing numbers
Money	Real Money	Solving problems involving basic operations of add, subtract, divide, and multiply
Fraction Circles	Construction Paper	Operations with fractions and decimals
Geoboards	Dot Paper	Developing conceptual understanding of area and perimeter- plane geometry
Algebra Tiles	Block, Buttons, Coins	Modeling of integers and variables with basic operations
Equation Mats	Construction Paper	Manipulating single-variable equations

Appendix B

Teaching Standards

Standard 1: Worthwhile Mathematics Tasks

Standard 2: Teacher's Role in Discourse

Standard 3: Students' Role in Discourse

Standard 4: Tools for Enriching Discourse

Standard 5: Learning Environment

Standard 6: Analysis of Teaching and Learning

Visit *www.nctm.org* for details of the teaching standards. Please note that some materials in the NCTM website can only be accessed by fee paying members. You are encouraged to join the NCTM so as to have access to valuable teaching resources.

References

Ames, C. (1992). Classroom: Goals structures and student motivation. *Journal of Educational Psychology, 84*(3), 261-271.

Asquith, P., Stephens, A., Knuth, E.J., & Alibali, M.A. (2007). Middle school mathematics teachers' knowledge of students' understanding of core algebraic concepts: Equal sign and variable. *Mathematical Thinking and Learning, 9*(3), 249-272.

Ausubel, D. (1963). *The psychology of meaningful verbal learning.* New York: Grune & Stratton.

Baker, S.K., Simmons, D.C., & Kameenui, E.J. (1994). Making information more for students with learning disabilities through design of instructional tools. *LD Forum, 19*(3), 14-18.

Baroody, A. (1999). Children's relational knowledge of addition and subtraction. *Cognition and Instruction, 17*(2), 137-175.

Behr, M., Khoury, H., Harel, G., Post, T., & Lesh, R. (1997). Conceptual units analysis of preservice elementary school teachers' strategies on a rational-number-as-operator task. *Journal of Research in Mathematics Education, 28*(1), 48-60.

Behr, M., Lesh, R., Post, T., & Silver, E. (1983). Rational number concepts. In R. Lesh (Ed.), *Acquisition of mathematics concepts and processes* (pp. 91-126). New York: Academic Press.

Bezuk, N., & Cramer, K. (1989). Teaching about fractions: What, when, and how? In P. Trafton (Ed*.), National Council of Teachers of Mathematics 1989 Yearbook: New Directions for Elementary School Mathematics* (pp. 156-167). Reston, VA: NCTM.

Bohan, H. (1990). Mathematical connections: Free rides for kids. *Arithmetic Teacher, 38*(3), 10-14.

Booth, L.R. (1989). A question of structure. In S. Wagner and C. Kieran (Eds.), *Research issues in the learning and teaching of algebra.* Reston, VA: NCTM.

Brekke, G. (1996). A decimal number is a pair of whole numbers. In L. Puig & A. Gutierrez (Eds.), *Proceedings of the 20 conference for the International Group for the Psychology of Mathematics Education* (Vol. 2, pp. 137 – 144). Valencia, Spain: PME.

Britt, M.S., Irwin, K.C., Ellis, J., & Ritchie, G. (1993). *Teachers raising achievement in mathematics: Report to the Ministry of Education.* Auckland, New Zealand: Auckland College of Education.

Brown, A. (2002). Patterns of thought and prime factorization. In S. Campbell & R. Zazkis (Eds.), *Learning and teaching number theory: Research in cognition and instruction* (pp. 131-137). Westport: Ablex Publishing.

Brown, A., Thomas, K., & Tolias, G. (2002). Conceptions of divisibility: Success and understanding. In S. Campbell & R. Zazkis (Eds.), *Learning and teaching number theory: Research in cognition and instruction* (pp. 41-82). Westport: Ablex Publishing.

Burns, M. (2009). Win- win math games. *CMC Communicator, 34*(1), 29- 32. Carnine, D. (1994). Introduction to the mini series: Diverse learners and prevailing, emerging, and research-based educational approaches and their tools. *School Psychology Review, 23*, 406-427.

Carnine, D., Slibert, J., & Kameenui, E.J. (1997). *Direct instruction reading* (3rd ed.). Upper Saddle River, NJ: Prentice Hall.

Carpenter, T.P. (1989). Teaching problem solving, In R.I. Charles & E.A. Silver, (Eds.), *The teaching and assessing of mathematical problem solving,* (pp. 187-202). Reston, VA: NCTM.

Carpenter, T.P., Franke, M.L., & Levi, L. (2003). *Thinking mathematically: Integrating arithmetic and algebra in elementary school.* Portsmouth, NH: Heinemann.

Carpenter, T.P., & Levi, L. (2000). *Developing conceptions of algebraic reasoning in the primary grades.* Wisconsin Center for Educational Research.

Chard, D. (2003). *Vocabulary strategies for the mathematics classroom.* Houghton Mifflin Math.

Christiansen, B., & Walther, G. (1986). Task and activity. In B. Christiansen, A.G. Howson, & M. Otte (Eds.), *Perspectives on mathematics education* (pp. 243-307). Holland: D. Reidel.

Clements, D.H., & McMillen, S. (1996). Rethinking concrete manipulatives. *Teaching Children Mathematics, 2*(5), 270-279.

Cramer, K.A., Post, T.R., & delMas, R.C. (2002). Initial fraction learning by fourth- and fifth grade students: A comparison of the effects of using commercial curricula with the effects of using the Rational Number Project curriculum. *Journal of Research in Mathematics Education, 33*(2), 111-144.

Demby, A. (1997). Algebraic procedures used by 13-to-15-year-olds. *Educational Studies in Mathematics, 33*(1), 45-70.

Dorward, J., & Heal, R. (1999). National library of virtual manipulatives for elementary and middle level mathematics. *Proceedings of WebNet99 World Conference on the WWW and Internet* (pp. 1510 to 1512). Honolulu, Hawaii Association for the Advancement of Computing Education.

Dreyfus, T. (1991). Advanced mathematical thinking process. In D. Tall (Ed.), *Advanced mathematical thinking* (pp. 25-41). Dordrecht, The Netherlands: Kluwer.

References

Ducan, K., Goldfinch, J., & Jackman, S. (1996). Conference review of 7th International Conference on the Teaching of Mathematical Modeling and Applications. *International Reviews on Mathematical Education, 28*(2), 67-69.

Elstak, I. E. (2007). *College students' understanding of rational exponents: A teaching experiment.* PhD Dissertation: Ohio State University.

English, L.D., & Sharry, P. (1996). Analogical reasoning and the development of algebraic abstraction. *Educational Studies in Mathematics, 30*, 135-157.

Falkner, K., Levi, L., & Carpenter, T. (1999). Children's understanding of equality: A foundation for algebra. *Teaching Children Mathematics, 12*, 232-236

Ferrari, P.L. (2002). Understanding elementary number theory at the undergraduate level: A semiotic approach. In S. Campbell & R. Zazkis (Eds.), *Learning and teaching number theory: Research in cognition and instruction* (pp. 97-115). Westport: Ablex Publishing.

Filloy, E., & Sutherland, R. (1996). Designing curricula for teaching and learning algebra. In A. Bishop, K. Clements, C. Keitel, J. Kilpatrick, & C. Laborde (Eds.), *International handbook of mathematics education* (Vol. 1, pp. 139-160). Dordrecht: Kluwer Academic.

Filloy, E., Rojano, T., & Solares, A. (2010). Problems dealing with unknown quantities and two different levels of representing unknowns. *Journal for Research in Mathematics Education, 41*(1), 52-80.

Fredericks, J.A., Blumfield, P.C., & Paris, A.H. (2004). School engagement: Potential of the concept, state of the evidence. *Review of Educational Research, 74*(1), 59-110.

Fuson, K.C., Wearn, D., Hiebert, J.C., Murray, H.G., Human, P.G., Olivier, A.I., Carpenter, T.P., & Fennema, E. (2001). Children's conceptual structures for multidigit numbers and methods of multidigit addition and subtraction. *Journal for Research in Mathematics Education, 28*(2), 130 – 162.

Gagnon, J., & Maccini, P. (2000). Best practices for teaching mathematics to secondary students with special needs: Implications from teacher perceptions and a review of the literature. *Focus on Exceptional Children, 32*(5), 1-22.

Graeber, A., & Johnson, M. (1991). *Insights into secondary school students' understanding of mathematics.* College Park, University of Maryland, MD.

Gray, E., & Tall, D. (1993). Success and failure in mathematics: The flexible meaning of symbols as process and precept. *Mathematics Teaching, 142*, 6-10.

Horton, S., Lovitt, T., & Bergerud, D. (1990). The effectiveness of graphic organizers for three classifications of secondary students in content area classes. *Journal of Learning Disabilities, 23*(1), 12-22.

Hughes, M. (1981). Can pre-school children add and subtract? *Educational Psychology, 1*, 207 – 219.

Hunker, D. (1998). Letting fraction algorithms emerge through problem solving. In L. J. Morrow (Ed.), *The teaching and learning of algorithms in school mathematics* (pp. 170-182). Reston, Virginia: National Council of Teachers of Mathematics.

Irwin, K.C. (2001). Using everyday knowledge of decimals to enhance understanding. *Journal for Research in Mathematics Education, 32*(4), 399 – 420.

Irwin, K.C., & Britt, M.S. (2004). Operating with decimal fractions as a part-whole concept. In I. Putt, R. Faragher, & M. Mclean (Eds.), *Mathematics Education for the Third Millennium: Towards 2010* (Proceedings of the 27th Annual Conference of the Mathematics Education Research Group, Australia, Townsville), Pp. 312 – 319.

Jitendra, A. (2002). An exploratory study of schema-based word-problem-solving instruction for middle school students with learning disabilities: An emphasis on conceptual and procedural knowledge. *The Journal of Special Education, 36*(1), 23-28.

Johnson, R. (2008). The sucker bet: Understanding and correcting students' misconceptions about mathematics. *CMC Communicator, 33*(1), 26-30.

Joram, E., Gabriele, A., Bertheau, M., Gelman, R., & Subrahmanyam, K. (2005). Children'suse of the reference point strategy for measurement estimation. *Journal for Research in Mathematics Education, 36*(1), 4-23.

Kaput, J., & Blanton, M. (2001). Algebrafying the elementary mathematics experience. In H. Chick, K. Stacey, J. Vincent, & J. Vincent (Eds.), *The future of the teaching and learning of algebra. Proceedings of the 12 th ICMI study conference* (Vol. 1, pp. 344-352).Dordrecht, The Netherlands: Kluwer Academic.

Kaput, J., Carraher, D., & Blanton, M. (2007). *Algebra in the early grades*. Mahwah, NJ: Erlbaum.

Karplus, R., Stephen, P., & Stage, E. (1983). Proportional reasoning of early adolescents. In R. Lesh & M. Landan (Eds.), *Acquisition of mathematical concepts and processes*. Orlando, FL: Academic Press.

Ke, F. (2008). A case study of computer gaming for math: Engaged learning from GamePlay? *Computers and Education, 51*(4), 1609-16210.

Kelly, K. (2002). Lesson study: Can Japanese methods translate to United States schools? *Harvard Education Letter* (May/June 2002), pp. 22-25.

Kieran, C. (1989). The early of algebra: A structural perspective. In S. Wagner and C. Kieran (Eds.), *Research issues in the learning and teaching of algebra*. Reston, VA: NCTM.

Kieran, C. (1992). The learning of algebra in schools. In D. Grouws (Ed.), *Handbook of research on mathematics teaching and learning,* (pp. 390-419). New York: MacMillan Publishing Company.

Kieran, C. (2006). Research on the learning and teaching of algebra. In A. Gutierrez & P. Boero (Eds.), *Handbook of Research on the Psychology of Mathematics Education: Past, Present and Future* (pp. 11-49). Sense Publishers.

Knuth, E., Stephens, A., McNeil, N., & Alibali, M. (2006). Does understanding the equal sign matter? Evidence from solving equations. *Journal for Research in Mathematics Education, 37*(4), 297-312.

Krishner, D. (1989). The visual syntax of algebra. *Journal for Research in Mathematics Education, 20*(3), 174-187.

Lamon, S.J. (1999). *Teaching fractions and ratios for understanding*. Mahwah, NJ: Lawrence Erlbaum Associates.

Lasik, E.V., & Siegler, R.S. (2007). Is 27 a big number? Correlational and causal connections among numerical categorizations, number line estimation, and numerical magnitude comparisons. *Child Development, 78*(6), 1723-1743.

Lee, L. (1996). An initiation into algebraic culture through generalization activities. In N. Bednarz, C. Kieran, & L. Lee (Eds.), *Approaches to algebra: perspective for research and teaching* (pp. 87-106). Dordrecht, The Netherlands: Kluwer.

Lee, L., & Wheeler, D. (1989). The arithmetic connection. *Education Studies in Mathematics, 20*, 41-54.

Lenz, B.K., Deshler, D.D., & Kissam, B. (2004). *Teaching content to all: Evidenced-based inclusive practices in middle and secondary schools.* Boston: Pearson Education, Inc.

Lo, J.J., & Watanabe, (1997). Developing ratio and proportion schemes: A story of fifth grader. *Journal for Research of Mathematics Education, 28* (2), 216 – 236.

Lochhead, J. (1991). Making math mean. In E. von Glasersfeld (Ed.), *Radical constructivism in mathematics education* (pp. 75-87). Dordrecht, The Netherlands: Kluwer Academic Publishers.

Ma, L. (1999). *Knowing and teaching elementary mathematics: Teachers' understanding of fundamental mathematics in China and the United States.* Mahwah, NJ: Lawrence Erlbaum.

Maccini, P., & Ruhl, K. (2000). Effects of a gradual instructional sequence on the algebraic subtraction of integers by secondary students with learning disabilities. *Education and Treatment of Children, 23*, 465-489.

Martinez, J. (1988). Helping students understand factors and terms. *Mathematics Teacher, 81*, 747 – 751.

Matz, M. (1980). Towards a computational theory of algebraic competence. *Journal of Mathematical Behavior, 3*(1), 93-166.

Matz, M. (1982). Towards a process model for school algebra errors. In D. Sleeman & J. Brown (Eds.), *Intelligent tutoring systems* (pp. 25-50). New York: Academic Press.

Mercer, C.D., Lane, H.B., Jordan, L., Allsopp, D.H., & Eisele, M.R. (1996). Empowering teachers and students with instructional choices in inclusive settings. *Remedial and Special Education, 17,* 226-236.

Michaelidou, N., Gagatsis, A., & Pitta-Pantazi, D. (2004). The number line as a representation of decimal numbers: A research with sixth grade students. Proceedings of the 28th Conference of the *International Group for the Psychology of Mathematics Education*, Vol. 3 (pp. 305-312).

Millhiser, W. (2011). The value of open-ended questions. *Issues in Education, 38*(2), 5-10.

Moss, J., & Case, R. (1999). Developing children's understanding of the rational numbers: A new model and an experimental curriculum. *Journal for Research in Mathematics Education, 30*(2), 122-47.

Moyer, P.S., Bolyard, J.J., & Spikell, M.A. (2002). What are virtual manipulatives? *Teaching Children Mathematics, 8*(6), 372-377.

National Council of Teachers of Mathematics (NCTM) (2000). *Principles and standards for school mathematics.* Reston, VA: Author.

Ojose, B. (2008). Applying Piaget's theory of cognitive development to mathematics instruction. *The Mathematics Educator, 18*(1), 26-30.

Ojose, B. (2010). *Mathematics education: Perspectives on issues and methods of instruction.* Dubuque, IA: Kendall Hunt Publishing.

Ojose, B. (2012). Subject matter knowledge and instructional processes: What is the connection with regard to teaching algebraic concepts? *Proceedings of the Annual Conference of the American Educational Research Association (AERA)*. Vancouver, British Columbia, Canada (April 13-17, 2012).

Okebukola, P. (1992). Can good concept mappers be good problem solvers in science? *Educational Psychology, 12*(2) 113-130.

Ozmantar, M.F. (2005). *An investigation of the formation of mathematical abstractions through scaffolding*. PhD Thesis, University of Leeds.

Pearn, C., & Stephens, M. (2004). Why do you have to probe to discover what Year 8 students really think about fractions? In I. Putt, R. Faragher, & M. McLean (Eds.), *Mathematics education for the third millennium: Towards 2010. Proceedings of the 27th Annual Conference of the Mathematics Education Research Group of Australasia*. Townsville, 27-30 June. (pp. 430-437).

Perry, M. (1991). Learning and transfer: Instructional conditions and conceptual change. *Cognitive Development, 6*, 449-468.

Raftopoulos, A. (2002). The spatial intuition of number line and the number line. *Mediterranean Journal for Research in Mathematics Education, 1*(2), 17-36.

Raphael, D., & Wahlstom, M. (1989). The influence of instructional aids on mathematics achievement. *The Journal of Research in Mathematics Education, 20*(2), 173-190.

Resnick, L.B., Bill, V.L., Lesgold, S.B., & Leer, M.L. (1991). Thinking in arithmetic class. In B. Means, C. Chelemer, & M.S. Knapp (Eds.), *Teaching advanced skills to at –risk students* (pp. 27-53). Jossey-Bass: San Fransisco.

Rittle-Johnson, B., & Alibali, M.W. (1999). Conceptual and procedural knowledge of mathematics: Does one lead to the other? *Journal of Educational Psychology, 91*(1), 175-189.

Roach, D., Gibson, D., & Weber, K. (2004). Why is root 25 not ±5? *Mathematics Teacher, 97*(1), 5-10.

Sfard, A. (1991). On the dual nature of mathematical conceptions: Reflections on processes and objects as different sides of the same coin. *Educational Studies in Mathematics, 22*, 1 – 36.

Siegler, R.S. (1991). In young children's counting, procedures precede principles. *Educational Psychology Review, 3*, 127-135.

Siegler, R.S. (1995). How does change occur? A microgenetic study of number conservation.*Cognitive Psychology, 28*, 225-273.

Siegler, R.S. & Booth, J.L. (2004). Developing numerical estimation in young children. *Child Development, 75*(2), 428-444.

Simon, H.A. (1981). *The sciences of the artificial*. Cambridge, MA: The MIT Press.

Singh, P. (2000). Understanding the concepts of proportion and ratio among grade nine students in Malaysia. *International Journal of Mathematical Education in Science and Technology, 31*(4), 57999.

Sneed, D., & Snead, W.L. (2004). Concept mapping and science achievement of middle grade students. *Journal of Research in Childhood Education, 18*(4), 306-

Starkey, P., & Gelman, R. (1982). The development of addition and subtraction abilities prior to formal schooling in arithmetic. In T. Carpenter, J.M. Moser, & T.A.

Romberg (Eds.), *Addition and subtraction: A cognitive perspective* (pp. 99 – 116). Hillsdale, NJ: Erlbaum.

Steinle, V. (2004). *Detection and remediation of decimal misconceptions*. Unpublished PhD thesis, University of Melbourne, Melbourne.

Steinle, V. (2005). The incidence of misconception of decimals among students in grades 5 to 10. In C. Kanes, M. Goos, & E. Warren (Eds.), *Teaching mathematics in new times* (Vol. 2, 548-555). Brisbane: MERGA.

Sullivan, P., Clarke, D., & Clarke, B. (2009). Converting mathematical task to learning opportunities: An important aspect of knowledge for mathematics teaching. *Mathematics Education Research Journal, 21*(1), 85-105.

Swafford, J.O. & Langrall, C.W. (2000). Grade 6 students' preinstructional use of equations to describe and represent problem situations. *Journal of Research in Mathematics Education, 31*(1), 89-112.

Sweller, J., Mawer, R.F., & Ward, M.R. (1983). Development of expertise in mathematical problem solving. *Journal of Experimental Psychology: General, 112*, 639-661.

Tall, D., & Thomas, M. (1991). Encouraging versatile thinking in algebra using the computer. *Educational Studies in Mathematics, 22*, 125-147.

Thanheiser, A. (2009). Preservice elementary school teachers' conceptions of multidigit whole numbers. *Journal for Research in Mathematics Education, 40*(3), 251-281.

Thomson, S., & Walker, V. (1996). Connecting decimals and other mathematical content. *Teaching Children Mathematics, 8*(2), 496-502.

Trigueros, M., Oktac, A., & Manzanero, L. (2007). Understanding of systems of equations in linear algebra. *CERME, 5*, 2359-2368.

Vaidva, S. R. (2004). Understanding dyscalculia for teaching. *Education, 124*(4), 717.

Van de Walle, J. (2004) *Elementary and middle school mathematics: Teaching developmentally* (5th ed.). Boston, MA: Allyn & Bacon.

Van de Walle, J. (2007) *Elementary and middle school mathematics: Teaching developmentally* (6th ed.). Boston, MA: Allyn & Bacon.

Vermeulen, N., Olivier, A., & Human, P. (1996). Students awareness of the distributive property. *International Conference for the Psychology of Mathematics Education (PME 20)*. Valencia, Spain.

Ward, R.A., (2005). Using children's literature to inspire K-8 preservice teachers' future mathematics pedagogy. *The Reading Teacher, 59*(2), 132-143.

Warren, E., & Cooper, T. (2001). Theory and practice: Developing an algebra syllabus for P-7. In H. Chick, K. Stacey, J. Vincent, & J. Vincent (Eds.), *The future of the teaching and learning of algebra. Proceedings of the 12 th ICMI study conference* (Vol. 1, pp. 641-648). Dordrecht, The Netherlands: Kluwer Academic.

Weller, K., Clark, J., Dubinsky, E., Loch, S., McDonald, M., & Merkovsky, R. (2003). Student performance and attitudes in courses based on APOS Theory and the ACE Teaching Cycle. In A. Selden, E. Dubinsky, G. Harel, & F. Hitts (Eds.), *Research in Collegiate Mathematics Education V* (pp. 97-131). Province, RI: American Mathematical Society.

Whyte, J.C., & Bull, R. (2008). Number games, magnitude representation, and basic skills in preschoolers. *Developmental Psychology, 44*(2), 588-596.

Willerman, M., & Mac Harc, R.A. (1991). The concept map as an advanced organizer. *Journal of Research in Science Teaching, 28*(8), 705-712.

Witzel, B.S., Mercer, C.D., & Miller, D. (2003). Teaching algebra to students with learning difficulties: An investigation of explicit model. *Learning Disabilities Research & Practice, 18*(2), 121-131.

Wood, T., & Sellers, P. (1996). Deepening the analysis: Longitudinal assessment of a problem-centered mathematics program. *Journal for Research in Mathematics Education, 28*, (2), 163-186.

Woodward, J., Baxter, J., & Howard, L. (1994). The misconceptions of youth: Errors in their mathematical meaning. *Exceptional Children, 61*(2), 126-136.

Wu, H. (2007). *Order of operations and other oddities in school mathematics*. Available online: http://math.berkeley.edu/~wu/.

About the Author

Dr. Bobby Ojose is currently an Assistant Professor of mathematics education at the Youngstown State University, Ohio. He started his college level teaching career at the University of Redlands, California, where he taught courses in general education and mathematics education. He also was instrumental to the MA and doctoral programs at the university because of his involvement in teaching the quantitative research methods courses for those programs. At the Youngstown State University, Dr. Ojose is teaching methods courses in mathematics for the early, middle, and AYA programs. His research interests encompass mathematics education and teachers' content and pedagogical knowledge. His work also involves the use of technology in supporting mathematics teaching and learning. Dr. Ojose obtained his doctorate from the University of Southern California, Los Angeles.

Printed in Great Britain
by Amazon